Atomic Bill

Atomic Bill

A Journalist's Dangerous Ambition in the Shadow of the Bomb

Vincent Kiernan

Three Hills
an imprint of
Cornell University Press
Ithaca and London

Copyright © 2022 by Cornell University

All rights reserved. Except for brief quotations in a review, this book, or parts thereof, must not be reproduced in any form without permission in writing from the publisher. For information, address Cornell University Press, Sage House, 512 East State Street, Ithaca, New York 14850. Visit our website at cornellpress.cornell.edu.

First published 2022 by Cornell University Press

Printed in the United States of America

Library of Congress Cataloging-in-Publication Data
Names: Kiernan, Vincent, author.
Title: Atomic Bill : a journalist's dangerous ambition in the shadow of the bomb / Vincent Kiernan.
Description: Ithaca [New York] : Three Hills, an imprint of Cornell University Press, 2022. | Includes bibliographical references and index.
Identifiers: LCCN 2022009164 (print) | LCCN 2022009165 (ebook) | ISBN 9781501765636 (hardcover) | ISBN 9781501766008 (pdf) | ISBN 9781501766015 (epub)
Subjects: LCSH: Laurence, William L., 1888–1977. | Manhattan Project (U.S.)—Press coverage—United States. | New York times. | Science journalism—Moral and ethical aspects—United States—History—20th century. | Science news—Moral and ethical aspects—United States—History—20th century. | Journalists—New York (State)—New York—Biography. | Nuclear weapons information, American—History—20th century. | Journalists—Professional ethics—New York (State)—New York. | Atomic bomb—United States—History.
Classification: LCC Q225.2.U6 K54 2022 (print) | LCC Q225.2.U6 (ebook) | DDC 070.4/495092 [B]—dc23/eng20220528
LC record available at https://lccn.loc.gov/2022009164
LC ebook record available at https://lccn.loc.gov/2022009165

For Terri, Emily, and Matthew

Contents

Introduction

A Moth to the Flame

In just less than one year, an extravaganza of science and technology would be opening on a former garbage dump in Queens, New York: the 1964–65 New York World's Fair, expected to attract 55 million visitors with a gleaming vision of the future. For years, William Leonard Laurence, the famed science writer for the *New York Times*, had been quietly paid by fair president Robert Moses to help plan and promote the fair even as Laurence still worked for the paper—a glaring ethical violation for Laurence. But on that day in April 1963 Laurence committed an even worse infraction: while the paper's editorial-page editor was out of town, Laurence had the paper run an editorial calling on New York City to fund a permanent science museum at the fairgrounds—and proudly shared the editorial with Moses before it appeared in print. The next day, Laurence doubled down on his ethical trespass by testifying before a city appropriating board in favor of the expenditure. It was all of a piece for Laurence, who throughout his storied career, covering the atomic bombings of Japan as well as many other landmark stories in science and technology, ignored ethical and professional rules in order to satisfy his thirst for status and money and his desire to cozy up to the powerful and rich.

Journalists like Laurence have always been society's chroniclers of the moment—authors, in the phrase of uncertain provenance, of "the first rough draft of history." During the Civil War, reporters trekked along with the battling Union and Confederate armies and raced to get their dispatches to their newspapers by telegraph. At the turn of the nineteenth century, muckraking journalists horrified and disgusted readers with stories of filthy meatpacking plants and corrupt politicians, triggering crusades to clean up both. Today journalists sit in meeting rooms or watch streaming video feeds to keep tabs on the work of city councils and congressional committees.

For decades these authors of history's rough draft were usually uneducated scribblers, with no special expertise in the specific stories or topics they covered. Often moving together in a pack, these reporters from various publications scrambled from one story to the next, like the band of jabbering reporters memorialized in the 1928 Broadway play and 1931 film *The Front Page*. These reporters needed no particular expertise because their work had no particular focus: one day a reporter might cover a downtown fire; the next day, a municipal scandal. The work was all about the big story of the day, whatever it might be, and the reporters often seemed largely interchangeable with one another.

That was just fine with their editors and publishers, who sought to publish newspapers and magazines that appealed to the largest audience possible. Although occasional journalistic stars would emerge, reporting by and large was a fairly anonymous, blue-collar job with little job security—reporters knew that scores of would-be reporters would jockey after every position and were perfectly qualified (or as unqualified as anyone else) for it.

Around the start of the twentieth century, the modernizing of society began to wreak ripple effects on this journalistic ecosystem in many ways. One was the fact that the emergence of modern science created new products and technologies that touched the lives of Americans—electric lighting, aviation, early physics, the beginnings of modern medicine. During World War I, technology was a terrible and potent sword in the hands of both the Allies and the Central Powers. Chemical warfare, radio, and machine guns were among the technological advances that soldiers and their families back home came to know about. Editors and publishers began to perceive a public hunger for news about science, medicine, and technology—and with that hunger, a market opportunity, if only the publishers could employ reporters with the skills needed to find and report those stories. Gradually, a new generation of reporters focusing on science, medicine, and technology began to emerge. For example, Herbert B. Nichols was named natural science editor of the *Christian Science Monitor* in 1928, and Thomas R. Henry became science writer for the *Washington Star* in 1929. *New York Times* reporter Alva Johnston covered the 1922 meeting of the American Association for the Advancement of Science in Cambridge, Massachusetts, for which he received a Pulitzer Prize.[1]

It was the beginning of an essential era in journalism. In light of the degree to which today's society is drenched in science and technology, journalism and journalists truly would have been guilty of professional malpractice if they had not developed the ability to identify, interpret, and report on developments in science and technology. But with that task came great responsibility, a responsibility that journalists have not always satisfied.

Physics, for example, was one of the most exciting fields of science at the start of the twentieth century. The structure of the atom, not to mention its potential for releasing enormous quantities of energy, was completely shrouded in mystery. While schoolchildren today can relate the basics of the atom, with the nucleus at its center and electrons spinning around it like planets orbiting the sun, the world's top scientists spent the early decades after 1900 debating and refining what exactly was happening in the atom. Meanwhile, soon after 1900, Albert Einstein formulated special relativity and general relativity, two theories that set theoretical physics and astronomy back on their heels.

At this time, Laurence was an up-and-coming aviation reporter at Joseph Pulitzer's *New York World*, a pioneer of yellow journalism, the extensive use of sensationalism and scandal to build readership. The forty-one-year-old had spent his early career in a zig-zag, combining time at Harvard University with a trial run in journalism at the *Boston American*, army service in World War I, and attainment of a law degree before finally resolving to enter journalism full time at the *World*. On New Year's Eve in 1929 he caught the assignment of covering a lecture by James MacKaye, a

Dartmouth professor who claimed he could disprove Einstein's theory of relativity. Although Laurence privately realized that MacKaye knew little about Einstein, Laurence plunged ahead with a story with the style that he would perfect at the *Times*, combining breathless enthusiasm for unreliable research with a didactic approach that would land his story on the front page. He made MacKaye look like a genius, stating that MacKaye had presented "a new theory of the universe pronounced by scientists and philosophers as epoch-making and far-reaching in its consequences, and differing radically from the Einstein theory of relativity." Then Laurence described the scientific reaction to the lecture: "After a moment of impressive silence as he finished reading, the staid academic halls resounded to the sound of hearty applause." On the following Sunday, Laurence followed up with a lengthy feature story describing MacKaye and his theories.[2]

Within months, the *Times* would hire Laurence, and he would promptly begin writing stories about atomic research. Encountering each other again and again at scientific conferences and events across the country, Laurence and science reporters from other newspapers and wire services would forge a new type of journalistic pack, one focused on the reporting of science and technology. They realized that being scooped by others in the pack would draw criticism from faraway editors—much more than the praise an editor would make for having a scoop—so they often would agree among themselves about what was newsworthy and what was not about a story, and when it could be released as news, so that they all knew what the others were going to do.

While Laurence covered many other science stories during his tenure at the *Times*, he always maintained a special fixation on the atom. Laurence grasped the implications, both military and economic, when researchers began to understand that each atom had enormous energies locked inside it. When a second world war began to look inevitable, Laurence would try to draw attention to worrisome Nazi work in atomic energy while also tracking America's efforts. All throughout, he retained a slavish obedience to authority—scientific or governmental—combined with an obsession for upward social mobility and a fascination with those in power.

These characteristics made Laurence the obvious candidate to work for the Manhattan Project once its leader, Lt. Gen. Leslie R. Groves, decided he needed someone to write his own rough first draft of history—a set of press releases that would lead the public to accept the atom bomb after it was used in Japan. Groves knew that he could count on Laurence to produce exactly what he wanted, dramatizing the bomb's power while downplaying or ignoring potentially controversial aspects such as the deadly radiation and radioactive fallout that the bomb would produce. Laurence would ride the publicity wave generated by the Manhattan Project as far and as long as he could, using it to leverage book deals and television and radio appearances, even as his expertise grew stale and out of date. The publicity lent Laurence a halo of expertise so intense that even the Soviet Union dispatched a spy to see what could be learned from him. But the halo was fake: Laurence, like many other journalists before and after him who have lost their way, reached decisions that were unethical or otherwise against the public interest, all in the name of promoting his own brand. Laurence had stopped serving his readers and the larger public, and was focused on serving himself.

This biography of Laurence considers the ethics and professionalism of his almost thirty-four-year career at the *Times*, where he became arguably the best-known science writer of his day. The book reexamines the well-trod controversy of Laurence's involvement in the Manhattan Project. But it also considers ethical lapses less generally known about Laurence, such as the repeated conflicts of interest represented by his work on the World's Fair; his use of military press releases as stories that he filed for the *Times* from South Pacific weapons tests; and his quixotic campaign to use the *Times* to convey legitimacy on a fringe scientist whose theories had been rejected by the scientific establishment. It also examines whether Laurence gave the *Times*—and, more importantly, its readers—full value for his reporting effort in light of his preoccupation with building his own brand and doing the bidding of the powerful in society.

The unsavory picture of Laurence that emerges from this narrative should give pause to journalists, public officials, and citizens everywhere. The proper functioning of liberal democracy requires citizens and their representatives in government to have access to accurate, unbiased information that they can use in reaching judgments and decisions. While even the fairest-minded journalists, as humans, frequently fall short of the standard of providing accurate and unbiased information to their readers, viewers, or listeners, Laurence's example illustrates the perils to society when journalists focus on their own glory, enrichment, and association with the rich and powerful rather than seeking fair and accurate information for society.

1

The Second Coming of Prometheus

Laurence loved being a science writer, doggedly covering scientific conferences and educating crusty editors about the realities of science news. Prolific and successful, within seven years he shared a Pulitzer Prize with science journalists from four other media organizations for coverage of a scientific conference marking the three hundredth anniversary of Harvard University's founding. In particular, the job enabled Laurence to indulge his fascination with atomic science, which long predated his arrival at the *Times*. Fortunately for Laurence and his readers, New York featured prominently in much atomic research, such as the Columbia University physicists led by John R. Dunning who found in 1935 that lithium atoms released energy when bombarded by neutrons. The process released 200 million times more energy than the amount of energy that was consumed in the process. Although physicists struggled to understand what was going on, Laurence and other science journalists immediately recognized that the phenomenon might have practical use. Laurence wrote that Dunning's research could be "the greatest advance so far toward the practical utilization of the inconceivably large quantities of energy known to be locked in the heart of atoms."[1] Laurence's story said nothing about military applications, but other insightful observers grasped that prospect. "Successful completion of these experiments, making the techniques commercially accessible, will make present power sources and methods as antiquated as the mastodon," reported the Catholic weekly *Commonweal*. "At the same time, if used destructively, it has been said by sober and reputable scientists, [the technology] may blow the known world to smithereens."[2]

A vital step forward came in December 1938, when researchers Lise Meitner, Otto Hahn, and Fritz Strassmann discovered the process of nuclear fission—splitting an atom. They fired neutrons at uranium, breaking apart some of the uranium atoms. The physicist Enrico Fermi effectively hijacked a physics conference at George Washington University in Washington, DC, on January 27, 1939, when he stood up and announced the development. Physicists at the conference were agog, and several dispatched news to their home laboratories to try to replicate the work. The first media coverage appeared the next day on the front page of the *Evening Star* in Washington, for which the physics conference was a local event. Thomas R. Henry, the paper's science writer, reported "the cracking of atoms with the release of the greatest energies ever known on earth." However, Henry did not suggest that the process could be used to create a weapon; indeed, he wrote that "the practical uses of the discovery remain vague."[3]

An Associated Press story by Howard Blakeslee about the accomplishment ran on the same day in many newspapers. On the West Coast, the physicist Luis Alvarez spotted Blakeslee's story in the *San Francisco Chronicle* during a haircut in the campus barber shop at the University of California at Berkeley. "I stopped the barber in mid-snip and ran all the way to the Radiation Laboratory to spread the word," Alvarez wrote in his memoirs.[4]

But the *Times* missed the story because Laurence had not attended the meeting. "I've never forgiven them for that, that they cheated me out of a story because I wasn't in Washington at that particular meeting," Laurence complained years later.[5] A day later the *Times* ran an Associated Press story about the energy release. In Germany, Hahn saw the *Times* article and added a handwritten annotation to a clipping, "Pure American exaggeration."[6]

The next important development happened a month later, and Laurence did not miss it. On February 24, 1939, when Fermi and the Danish physicist Niels Bohr spoke to a group of physicists at Columbia University, Laurence was the only reporter present, sitting in the audience next to Dunning. Fermi and Bohr discussed how using a neutron to split one atom could release an additional neutron that could split an additional atom, with that process occurring again and again, producing a runaway release of energy that they called a "chain reaction." As a lover of words, Laurence experienced an epiphany. "It took some time—it seems an eternity in retrospect—before the true meaning of those two simple words penetrated my consciousness. And then, suddenly, my conscious mind came to life and their meaning was revealed to me," he later wrote in his book *Men and Atoms*. "I remember saying to myself, 'This is the Second Coming of Prometheus, unbound at last after some half a million years, bringing down a fire from the original flame that had lighted the stars from the beginning.' "[7]

Laurence realized that if the Nazis developed atomic technology, the chain reaction could hand them a tremendously powerful weapon. After the meeting, Laurence prevailed on Dunning to introduce him to Fermi and Bohr so he could ask whether a small amount of the isotope uranium-235 could be used to create a bomb with the power of thousands of tons of TNT. "Fermi was the first to speak, after a pause long enough to give me the distinct feeling that I had asked a rather embarrassing question," Laurence wrote in *Men and Atoms*. "We must not jump to hasty conclusions," Fermi responded at last. "This is all so new. We will have to learn a lot more before we know the answers. It will take many years."[8]

O'Neill, the science journalist for the *New York Herald Tribune* and a friend and competitor of Laurence's, also began to realize the implications of what scientists were finding in the atom. After Dunning and the physicist Alfred O. C. Nier of the University of Minnesota published a letter about their atomic research in the prestigious physics journal *Physical Review* in September 1940—a letter that hid as much as it revealed about their work—John J. O'Neill deduced that the researchers had managed to separate atoms of uranium-235 from the more common atoms of uranium-238, a task essential for using the chain reaction to release energy. O'Neill stayed up all night writing a lengthy news story. He showed his draft to Dunning the next day, and they began a series of iterations in which Dunning and fellow Columbia physicist George B. Pegram would point out errors and O'Neill would produce a new draft of the story. Each step in the process "was twenty-four hours of torment for fear some other newspaper would penetrate Prof. Dunning's secret, and beat the writer on what was obviously the most important story ever published," O'Neill later recalled.[9] He wasn't scooped, but when O'Neill's story was ready, his editors killed it because it was too long and technical. O'Neill then approached the editors of *Harper's Magazine*, who agreed to publish it in the June 1940 issue, and O'Neill tipped off Laurence on the condition that Laurence hold his story until after *Harper's* hit the newsstands with O'Neill's story about May 20.

Instead, Laurence rushed his own story into print on May 5. Signifying the importance that his editors attached to it, the *Times* placed the story in the leftmost column of the front page. The fact that it ran in a Sunday edition only increased its visibility, readership, and impact. The headline, "Vast Power Source in Atomic Energy Opened by Science," accurately summarizes the main thrust of the story—that experiments by Dunning at Columbia had documented that triggering fission in uranium-235 could be an important new energy source. As little as five pounds of U-235 could release energy equivalent to 25–50 million pounds of coal or 15–30 million pounds of gasoline, Laurence wrote. He described research by Dunning and Nier to study the properties of U-235. Midway through the article, Laurence turned to the military potential of U-235, saying that scientists were reluctant to discuss their research because of "the tremendous implications that this discovery bears on the possible outcome of the European war." Laurence reported that the entire German scientific establishment had been set to the task of harnessing atomic science to the German war effort. He concluded the article by noting that scientists were reluctant to discuss methods of mass producing U-235. "As to this, scientists greet the questioner with a profound silence."[10]

Subsequently, Laurence and O'Neill squabbled over which of them had first reported the news that uranium-235 could be a source of immense power. After the war's end, in 1947, O'Neill wrote a letter to the editor of the *Saturday Review of Literature* claiming that he was first. In a reply letter to the magazine, Laurence retorted that what O'Neill wrote was not really news because it had all been reported a year earlier. "No wonder the editors of the *Herald Tribune*, with a world war on their hands, failed to stop the presses," Laurence sniffed. What made his own story newsworthy, Laurence claimed, was his revelation of extensive research by Nazis into developing an atomic weapon; O'Neill's story did not include that information. However, Laurence did not mention the Nazis until the eleventh paragraph of the story. The first ten paragraphs all focused on Dunning's work, as typified by the lead paragraph, "A natural substance found abundantly in many parts of the earth, now separated for the first time in pure form, has been found in pioneer experiments at the Physics Department of Columbia University to be capable of yielding such energy that one pound of it is equal in power output to 5,000,000 pounds of coal or 3,000,000 pounds of gasoline, it became known yesterday."[11] Indeed, in later years, Laurence himself described his 1940 story as being about nuclear research, not about the Nazis. A *Times* biography of him, which he approved, describes the story in this way: "In 1940, Mr. Laurence wrote the first comprehensive account in The Times of the meaning and the potentialities of the fission of uranium 235 and its possibilities as an atomic bomb and for power uses."[12]

In Nier's home state of Minnesota, two neighboring newspapers reacted very differently to Laurence's May 5 story. An editor at the *Minneapolis Tribune* noticed that a local professor (Nier) had played a role, so he assigned a reporter to run down that angle. But in an interview, Nier said little about the importance of the research, so the reporter wrote a bland story that was relegated to page 2, saying that the experiment amounted to "a partial revelation of the secret of atomic power" and cautioned that the notion of a uranium-fueled bomb "is exceedingly premature."[13] At the *Minneapolis Star-Journal*, however, the executive editor, Stuffy Walters, sensed that the story was big news, so he ran both a local story and the AP story and a picture of Nier across more than half of the front page, liberally using all capitals. "A University of Minnesota physicist has found a way to release in harness the terrific energy of the atom. . . . 'THE EFFECT OF USE OF COMMERCIALLY PRODUCED URANIUM ISOTOPE 235 IN WAR IS UNTHINKABLE,' DR. WILLIAMS SAID."[14]

Other journalists followed up on Laurence's story. *Time*, for example, stressed that the project was long term. "While it is known that months ago every German scientist in the field of atom-smashing was ordered to concentrate on uranium, it would take slightly less than 100,000 years to produce the necessary pound of the substance under present methods," the newsmagazine reported.[15]

Laurence had hoped that his story would galvanize the US government into starting a vibrant atom-bomb project to compete effectively with the German A-bomb efforts. To Laurence's disappointment, the only apparent reaction by the government to his *Times* article was from Sen. Sheridan Downey, a Democrat from California who inserted Laurence's article into the *Congressional Record.* But Downey did not use the article to issue the clarion call to action that Laurence sought, instead warning senators that atomic energy could threaten the economic viability of other energy industries such as petroleum and coal.[16]

Meanwhile, Laurence's *Times* colleague Waldemar Kaempffert sought to walk back any excitement that Laurence had created with his article. A week after Laurence's story, Kaempffert wrote that atomic power would not supplant traditional fuels. Uranium-235 was the only form of uranium usable for power, but scientists could suggest no practical way of separating large quantities of uranium-235 from other forms of uranium. "The prospect of using U-235 in the present war is zero," Kaempffert declared.[17]

Through much of Laurence's career, Kaempffert would assume this role of puncturing Laurence's balloon. The two had very different jobs and approaches to those jobs. Laurence was assigned to report and write news stories for the news pages of the paper; Kaempffert wrote unsigned editorials for the paper and bylined stories for the Sunday edition's Week in Review section. While Laurence was ever the dramatist in his writing, Kaempffert eschewed hyperbole, declaring in the journal *Science* in 1935 that "heaven forbid that the popularizer should rely too much on emotion. We have passed the stage when gasping wonder can pass for popularization."[18]

With neither in charge of the other, they often worked without coordination or even in open competition. Kaempffert often would run articles poking holes in whatever Laurence was trying to advance, mystifying the paper's editors and no doubt its readers as well. Turner Catledge, who was the paper's managing editor from 1951 to 1964 and executive editor from 1964 to 1968, put the situation succinctly in an oral history interview. "Bill didn't like Kaempffert. He didn't consider him to be quite a good scientist; Kaempffert, I think, sort of had his misgivings about Bill. There were jealousies."[19]

Henry A. Barton, the first director of the American Institute of Physics, worked closely with many science journalists of the time, including both Laurence and Kaempffert. "The interesting thing is that there was no connection inside the Times organization between those two men," Barton recalled in 1970. "Kaempffert, the Sunday man, never paid any attention to the daily man and vice versa. Separate files [*sic*]; they never talked to each other as far as I knew. I don't mean that they were unfriendly or anything, but I don't think they ever cooperated or helped each other with stuff."[20]

Barton recalled that Kaempffert covered few scientific meetings. "Kaempffert was the German encyclopedic type. He was more likely to talk than to listen actually, but he read; he read a great deal, Laurence went out and covered meetings. He was everywhere. And he asked questions, too." Barton recalled that Kaempffert, Laurence, and a few other journalists demanded advance copies of physics journals in the mid-1930s. Laurence "took me to school about that several times—[he wanted] everything, absolutely everything."[21]

Laurence was wrong about his story not being on the radar screens of Washington officials. The war was not going well for the Allies; the Germans had invaded Norway in April, and the British had failed to repel them, while invasions of Belgium, France, Luxembourg, and the Netherlands were imminent. Unknown to Laurence, his story had seized the attention of the Advisory Committee on Uranium, which President Franklin D. Roosevelt had formed a year earlier to investigate the military potential of uranium and recommend how the government should proceed with it. In a letter to Gen. Edwin Watson, secretary to President Roosevelt, the committee wrote, "The article of May 5th in the New York Times shows the widespread interest in this subject." But the committee said the article was no breach of security. "No reference was made to the particular work in which the Government is participating, nor was anything said in the article that is not generally known to physicists."[22]

Physicists in the know wrung their hands about Laurence's reporting. Nier, for example, tried to tamp down attention to the issue, telling the United Press that "the isolation of the isotope has little commercial or military value at present."[23] Columbia's Pegram also was vexed. "The newspaper outbursts about uranium energy seem to be beyond control," he complained to Lyman J. Briggs, who headed the uranium committee.

"I suppose they do little harm. Certainly they tell nothing that physicists here and elsewhere are not well acquainted with and nothing that they have not been talking about for the past year, but still I do not like it."[24] Eight days later, Pegram also complained to the physicist Hans Bethe that "W. L. Laurence got away with stuff that was quite distasteful to all of us at Columbia," though the physicist allowed that Kaempffert's editorial "did a rather good job of cancelling any excitement that might have been produced by Laurence's article."[25]

Dunning wrote Nier that he had failed to convince Laurence not to write the story. "We had succeeded in stopping all previous attempts to write stories, and thought that all was quieted down. However, Laurence and the Times thought they had a big story, and insisted on printing it without any sanction on our part whatever, on the grounds that this was still a free press." Dunning insisted to Nier that he had provided no information to Laurence. Pegram had talked with Laurence, Dunning reported, "and his major effort was in trying to keep Laurence calmed down and to write a conservative story or none at all."[26]

The article also came to the attention of military intelligence, which decided to track references to Laurence's article. For example, intelligence officials noted that a religious publisher cited Laurence's 1940 newspaper article in one of its publications. After a businessman from Evansville, Illinois, read the publication, he brought up the topic to Arthur Holly Compton, a physicist who happened to be involved in the Manhattan Project. Compton reported the conversation as a possible security breach.[27]

James A. Michener, who eventually would become a world-famous novelist but at the time was a visiting professor at Harvard, also was seized by Laurence's article. With its allusion to German atomic weapons programs, Laurence's piece "exploded in my mind like a flash of lightning," Michener wrote in a memoir more than fifty years later. "I hunched over the paper and read every word with extreme care." A naval reservist, Michener was called to active duty in 1944, and during his service he paid special attention to intelligence reports on German research. At one point, Michener decided that he wanted to reread Laurence's article. When he next visited New York, Michener strode into the New York Public Library in full uniform and asked a reference librarian for it. "Glaring at me with eyes popped wide open, he must have pressed a signal button, because I was quickly surrounded by two men who whisked me off to a private room for interrogation," Michener wrote. Eventually Michener convinced them that he was no spy, but the experience only reinforced Michener's suspicion that Laurence's reporting had struck on something important.[28]

The story of the army asking librarians to be on guard against Laurence's readers may seem far-fetched, but it would not be the only case of the government using libraries to restrict public access to information considered dangerous during World War II. In 1942 the American Library Association, on behalf of Secretary of War Henry Stimson, asked more than a hundred medium to large public libraries across the country to restrict their patrons' access to books on explosives, secret ink, and codes. Some librarians complied happily, while others complained that the plan failed to control information available at other sources such as other libraries and bookstores.[29]

Laurence's May 1940 article also triggered ripples within another government—that of the Soviet Union. When George Vernadsky, a history professor at Yale, saw the *Times* article, he sent a copy to his father, Vladimir Vernadskii, a geologist in the Soviet Union. Vernadskii noted Laurence's emphasis on uranium 235 and wrote to a colleague in the Soviet Academy of Sciences urging that a plan be developed to make sure that the Soviet Union would have adequate access to the isotope. He also wrote to a member of the Central Committee about the prospects of fission. One historian says the letter "appears to have been the first attempt by Soviet scientists to alert senior government scientists to the importance of nuclear fission."[30]

Laurence's story in the *Times* even attracted the attention of science fiction writers. Three months after the newspaper article appeared, *Astounding Science-Fiction* ran an article by its editor, John W. Campbell Jr. (writing under the pen name Arthur McCann), providing a primer on atomic energy. The article complained that Laurence's *Times* piece was too upbeat: "It sounded as though someone next door could, tomorrow, produce an atomic engine worth millions." By contrast, Campbell wrote, Kaempffert's article was "retracting the emphasis and enthusiasm, but none of the facts."[31] Laurence's article also served as key source material for the science fiction writer Cleve Cartmill, who wrote a fiction story, "Deadline," in March 1944 that seemed so well informed that army intelligence suspected that the author had access to inside information on the Manhattan Project.[32]

Frustrated by his apparent inability to goad the US government into research into atomic energy, Laurence submitted another article on the subject, this time to the *Saturday Evening Post*. The magazine's editor was nervous about Laurence's claims and required Laurence to provide confirmation from key scientists. No shrinking violet, Laurence requested an endorsement from the physicist Karl Taylor Compton, the president of the Massachusetts Institute of Technology.[33] Rather than review the article himself, Compton delegated the task to his former student Philip Morse, a physics professor at MIT. Morse produced a six-page, single-spaced critique of the draft, arguing that it suffered from hype and inaccuracy: "The difficulty is that most of Mr. Laurence's facts are correct; it is the atmosphere of the article which is wrong and which would annoy most scientific men." For example, Morse complained that the article gave too much credit to Columbia University researchers for the developments in U-235 and downplayed the contributions of other researchers such as Meitner, Hahn, and Nier. Morse warned Compton, "You can imagine the reaction on [*sic*] the rest of the workers in the field when they are told that their work has counted for nothing; you can imagine how anxious they will be to co-operate further with Columbia when they read this Columbia publicity."[34] Morse produced a somewhat shortened version of the critique, which John Rowlands, the institute's director of news services, sent Laurence.[35] Compton wrote Laurence separately, endorsing Morse's critique.[36] Laurence rewrote the article and sent the new version to Morse, thanking him for the comments. He wrote, "That is just the sort of criticism and help anyone in my position needs in order to do a good job in the difficult task of popularizing science. Such cooperation between the scientist and science reporter will, I am sure, serve a highly useful purpose for both science and public education."[37] Morse pronounced himself satisfied with the revised article and endorsed its publication.[38] The revised article also won the approval of the University of California, Berkeley physicist Edwin McMillan, who particularly praised Laurence for emphasizing the need for heavy radiation shielding around a reactor, which would make impractical most mobile uses of nuclear power.[39]

Published in September 1940, as the Allies worried that an invasion of Britain might be around the corner, the final version of Laurence's article dealt very little with the military prospects for atomic energy. Rather, it centered on a dramatic narrative account of some of the early history of atomic research, most notably the work of Meitner and Hahn, who had received only brief notice in Laurence's May 5 *Times* story. Laurence particularly focused on Meitner as a heroic and tragic figure who played a pivotal role in pointing physicists toward the track of research that eventually led to the development of atomic energy. Laurence dramatically related how Meitner was expelled by the Nazi regime for not being Aryan, despite her long ancestry in Germany. "Lise Meitner was on the train bound for Stockholm, sadly looking out of the window where she had spent her life devotedly in the pursuit of knowledge," Laurence wrote. "She was sixty years old, unmarried, and a woman without a country. She was going to a strange land, where she would try to resume her work." Laurence portrayed her as experiencing a flash of insight that the puzzling experimental results could be explained by postulating that uranium atoms could be split.[40] Once she arrived in Stockholm, Laurence wrote, Meitner communicated her thoughts and findings to other scientists, which "started off a set of events as dramatic as any in the history of man's endless quest for new means of mastery over his material environment."[41]

The *Post* story depicted Meitner much more dramatically than the *Times* article in May had. The *Times* article had omitted the train scene; it had mentioned her exile but did not attribute it to racial biases, much less a Jewish status.[42] And the *Post* article had factual errors. The eureka moment on the train that Laurence so movingly described didn't happen; the train journey transpired in July 1938, and the discovery of fission didn't happen until months later. Moreover, Otto Frisch was Meitner's nephew, not just her "friend," as Laurence stated.

Meitner was deeply unhappy with Laurence's portrayal, which implied she was Jewish. Meitner's biographer, Ruth Lewin Sime, wrote, "Lise was stung by the disregard for the truth, embarrassed among her colleagues, and unhappy to find herself embraced by the Jewish community, for which she felt no special affiliation." Sime called Laurence's article "a mix of gee-whiz science and docudrama" and held that Laurence's interpretation of the Meitner story strongly influenced later tellings of the story,[43] such as an article in *Time* that appeared shortly after the atomic bombings that described Meitner as "a Jewish woman scientist who had fled from Hitler's Reich to Copenhagen." American physicists "stood gallantly back" and allowed Meitner to confirm the phenomenon of atomic fission, *Time* said.[44] Sime also argues that an article about Meitner in

Current Biography was strongly influenced by Laurence's characterization of Meitner; indeed, that article quotes directly from the *Saturday Evening Post* article.[45] An article published in the *New York Herald Tribune* shortly after the atomic bombings described her as a "brown-eyed, gray-haired vivacious little Jewess" who was "embarrassed at all the attention she has attracted."[46] A week after the bombings, the *Washington Post* reprinted Laurence's *Saturday Evening Post* article, with the overline "Women without country provides important figure," which—though grammatically confused—was a clear reference to Meitner.[47]

Laurence's article moved on to describe work by Bohr, the experiments at Columbia University verifying the predictions, and the 1939 conference at George Washington University at which the results were announced. The key problem, Laurence wrote, was developing a way to extract U-235 because the methods then in existence would require more than 11 million years to yield one pound of the isotope. At this point, Laurence did raise the war, writing, "The tentacles of the swastika cast a shadow on the tranquil walls of our laboratories," because of the fear that better-equipped and better-staffed German laboratories might find a way to accelerate the extraction process. If so, the Germans would gain access to "the most powerful fuel ever dreamed of."[48]

In fact, Laurence specifically played down the prospect of an atomic bomb in his article. He instead portrayed nuclear energy as a power source for submarines and surface vessels. "One pound of pure U-235 would have the explosive power of 15,000 tons of T.N.T., or 300 carloads of 50 tons each. But such a substance would not likely be wasted on explosives," he wrote.[49]

Laurence larded the article with metaphors and dramatic flourishes. With the benefit of today's hindsight, one wonders if he would have been better off as a playwright. While an undergraduate student at Harvard, Laurence had taken a course in playwriting in which he befriended Eugene O'Neill; in both Boston and New York, he translated several plays, including two that were produced; and he was a lifelong member of the Dramatists' Guild. In the article, Laurence described the process of a chain reaction—in which a splitting atom emits neutrons that crash into other atoms, which themselves split and eject more neurons, perpetuating the process—as "a strange game of 'atomic golf' " in which the neutrons are the golf balls and atoms are the holes, and an atom that is struck by a neutron counts as a "hole in one." Laurence ended the *Saturday Evening Post* piece with a call to action for "industrialists and public-utility leaders" to financially support "the pioneer scientists in this highly important research." Notably, Laurence omitted government from the list of potential supporters. "It would be tragic indeed, if America were to lose the lead it is now believed to have in this field because its scientists, as the result of lack of funds, could not keep up in the race with their totalitarian rivals. A few thousands of dollars invested for research now may be worth hundreds of millions in the future."[50] Laurence again was trying to play the role of an atomic Paul Revere for his adopted nation, but he failed to trigger the same response that Revere had produced; consequently, Laurence would next seek to help the war effort in another fashion.

2

On the Army's Payroll

After the United States entered World War II in December 1941 following Japan's attack on Pearl Harbor, maintaining public support for the war effort was a key objective among American leaders. One way in which the US Army sought to shore up support on the home front was a little-known project by the army's Office of the Surgeon General to generate positive press coverage about steps the army was taking to protect the health of soldiers deployed overseas. Laurence's involvement in that project would presage his more widely known participation in the Manhattan Project later in the war.

The army project arose from a concern among military and medical officials in the United States that poor public understanding about military medicine might undermine public support for the war effort. Key among these were Morris Fishbein, MD, the editor of the *Journal of the American Medical Association*, who wielded enormous power throughout the medical community. During the war, Fishbein chaired the Committee on Information of the Division of Medical Sciences of the National Research

Council, an advisory organization headquartered in Washington. Over decades of working with reporters and the general press, Fishbein had developed strong opinions about the correct methods for disseminating information about medicine and science. He focused on providing access to information to favored journalists known to produce stores that he judged to be of high quality and orchestrating the release of those stories to produced maximum effect on the public. Before the war, Fishbein had been a prime mover in creating a system of advance access by selected journalists to articles being published in his journal. In return the journalists agreed not to publish their articles until a time set by Fishbein. This meant that when the articles finally were published, they created a burst of publicity about the research—and mentioned Fishbein's journal. This system, called the news embargo, has been copied by many journals and scientific associations and remains a key feature of how scientific and medical news is reported by journalists.[1]

Fishbein sought to adapt his techniques for the war effort even before the United States entered World War II. In late August 1940, the National Research Council's Committee on Information met in Washington, DC, to discuss medical communications. One topic was the creation of a committee regarding "general publicity for medicine" composed of journalists serving as consultants. The proposed members were Watson Davis of Science Service; Arthur T. Robb, editor of the newspaper trade publication *Editor and Publisher*; David Dietz of Scripps-Howard Newspapers; and Howard Blakeslee of the Associated Press. The four would serve without payment other than reimbursement of expenses.[2] The first three all agreed. "I am delighted to be of service in the present emergency," Dietz wrote.[3] But Blakeslee demurred. "I wish I could serve on the Advisory Committee, but it is not permissible for an Associated Press writer to serve in a publicity capacity," he responded to Lewis Weed, the physician who was chairman of the NRC's Division of Medical Sciences.[4]

To fill the hole on the committee left by Blakeslee, Morris Fishbein proposed adding Gobind Behari Lal of International News Service and Niles Trammell of the National Broadcasting Company. Weed objected that Lal's status as a naturalized US citizen might raise "wholly nonsensical" concerns, and Fishbein suggested Laurence as an alternative to Lal, overlooking the fact that Laurence also was not a native-born citizen. Fishbein told Weed that he preferred Laurence over Kaempffert because "I find it rather difficult to work with him." Davis agreed with Laurence's selection, calling Laurence "a thoroughly competent writer on scientific affairs."[5] The committee had its first meeting in New York on November 7, 1940.[6] Two days later, the committee's mission, charge, and journalistic members (including Laurence) were publicly announced in Fishbein's *Journal of the American Medical Association*.[7]

In December 1940 the committee met to discuss its mission. The journalists would be sent proof copies of journal articles and news tips from the military, and they would prepare releases that would be reviewed by Fishbein and then disseminated by the Federal Security Agency, a forerunner of today's Department of Health and Human Services. In the discussion, Laurence pressed for the development of documentary films, which he argued "would be a great help in telling the general public of the activities in medical defense."[8] Fishbein and the government would direct the journalists in the topics they covered and the messages the coverage advanced. The journalists were enthusiastic participants in this arrangement, as evidenced by Watson Davis's response when a medical journalist sought to be placed on the mailing list for press releases generated by the committee. "It ought to be a fair news source if plans underway are perfected," Dietz described the mailing list.[9]

The committee met periodically to discuss issues and develop strategies to address them. Their liaison to the NRC was Maj. Stephen McDonough, who previously had been AP's science writer in Washington, DC, and an early member of the National Association of Science Writers. That the participating journalists had surrendered their reportorial autonomy is clear from minutes of a June 1943 NRC meeting, which describe the journalists as taking direction from the committee regarding coverage of food rationing for invalids and hospitals. A general overview of the food distribution program "was assigned to Mr. David Dietz. To Mr. William Laurence was assigned a story dealing with the special conditions covered by the program, such as diabetes, tuberculosis, nephrosis, sprue, etc.," the minutes stated. When completed, the articles were to be sent to Fishbein, who would send them to the academy for distribution.[10] It is not clear whether Laurence ever delivered his story; no story like it was run by the *Times* under his byline. But it is clear that Fishbein and his colleagues were calling the shots.

A public relations crisis for the army regarding a yellow fever vaccine deepened the involvement by Laurence and other science journalists in supporting the government's efforts to shape public opinion. The army had administered more than 7 million doses of the vaccine to troops, most without incident, but more than 25,000 immunized service members in the western United States developed jaundice within two to five months after being immunized.[11] Most newspapers reacted mildly to the jaundice outbreak. The *Times*, for example, ran three short unbylined or wire-service stories; the only staff-written piece was a Sunday article by Kaempffert.[12] But the *Chicago Tribune* raised a hue and cry about the jaundice problem, running acerbic editorials questioning the competence of the program's management and whether mandatory immunization was prudent and calling for an investigation by an independent commission.[13] The controversy threatened to undermine public support for the war. Fishbein believed that poor public relations by the military were largely to blame for the controversy; investigation showed that the technical root of the problem with the vaccine was that certain batches incorporated blood serum from individuals who had had hepatitis B. Fishbein argued in a memorandum to senior military officials that "brief and intensive education of the public in regard to the nature of yellow fever, the virtues of the vaccine and the possibility of complications" followed by brief updates about the appearance of some cases of jaundice would have prevented the adverse public reaction.[14]

Fishbein proposed to bring the journalists on his NRC committee to the rescue. They "might come to Washington from time to time in order to make available for one day or for two days their ability and their information to the public relations divisions . . . aiding them in determining what material might be considered of special news value and also in preparing such material so that it would be usable by both the public and the medical press," he told the military officials. The journalists would not be paid other than reimbursement for expenses.[15]

"Regular writers and reporters would not do for there was no time to train them in the technicalities of medical terminology and the ramifications of clinical and military medicine," Robert Potter later wrote. "The Surgeon General knew the men he wanted but they were all over draft age. They had top posts in science writing on major newspapers and wire services, and were hardly prepared to backtrack down and into service as volunteers because of family obligations."[16]

But the arrangement also would put journalists in an ethical bind. Through their participation, they would have inside knowledge about the medical problems being faced by the US military and the military's attempts to solve them. They would have to keep that information under wraps until the designated time, and in fact they might never be able to use some of the information they had. On the issue of jaundice and yellow fever, the surgeon general's office was aware of the problem with the vaccine at least as early as April 1942 and had suppressed the information out of fear of undermining war morale.[17]

The details of the consulting arrangement were settled in late winter and early spring. In February 1944, Maj. Gen. Norman T. Kirk, the army's surgeon general, approached Laurence to see if he was interested in serving as a consultant to his office.[18] After getting Laurence's agreement, Kirk's deputy, Maj. Gen. George F. Lull, then asked Charles Merz, the city editor of the *Times*, for permission for Laurence "to spend a few days each month in the office of The Surgeon General," where he would be "preparing news releases on the work of the Medical Department." Lull appealed to the paper's willingness to uphold the war effort: "Such activity will be of considerable aid in improving the morale of the American people, by enabling them to understand how thoroughly the Medical Department is doing the job of preserving human life and, as far as possible, restoring men to normalcy."[19]

It would not be Laurence's first relationship with the US Army. In 1917, three months after the United States had entered World War I, Laurence had enlisted in the Army Signal Corps in part to escape from Harvard University, where he had been trying without success to graduate. Laurence had entered Harvard in 1908, virtually penniless and without family to support him. Nevertheless, his timing was propitious, because he was admitted at the tail end of the presidency of Charles W. Eliot, who changed the composition of the student body from principally the sons of the elite in the Northeast and New England to include a wider array of socioeconomic backgrounds and statuses.[20] Even so, Laurence's academic file at Harvard shows that he struggled at Harvard academically and financially; the university noted multiple complaints that he failed to repay loans from the university and individuals. Holds on his account repeatedly interrupted his studies. In September 1915 Laurence was arrested after an altercation with his roommate, Benjamin Stolberg, who after graduation would go on to become a journalist and labor activist. Laurence was found guilty of assault and battery, although he was released without having to spend any time in jail.[21] By May 1917 Laurence believed he was ready to graduate, but the university still blocked his account because of debts he owed, so he decided to try his hand with the army.

From the start, Laurence and the military were a poor match. He wanted to enlist in the Aviation Corps; because of his poor eyesight he cheated on the eye test by memorizing all eye charts used in the exam. When that approach failed, he turned to the Signal Corps. At Fort Devens in Massachusetts, Laurence was assigned to manual labor and caring for mules—one of which kicked him straight across the barn. He kept applying for other assignments that he thought would be better suited for him, such as a slot in officer's school or a post in intelligence, but his superiors consistently turned him down. Once Laurence was deployed to France, he helped operate telephone switchboards connecting French and American troops, an assignment that better capitalized on his fluency in multiple languages. But even then Laurence was a cantankerous soldier; for example, he decided to witness the entry of French troops into the city of Metz, even though US soldiers had been ordered to stay away from Metz so Americans would not distract from the French liberation of the city.

During World War II, almost three decades later, the army had a much higher view of Laurence as a civilian, coveting his journalistic skills to advance the war effort. By March 1944 the surgeon general's office had devised its plan for leveraging Laurence and the other science writers. The NRC would call a meeting of the committee each month, which would allow the journalists to travel to Washington, DC, to work in the surgeon general's office for at least two days at a time. That would not be their only contribution. "Some of the work would be done by the men in their own homes when the material is sent to them," Fishbein told Brig. Gen. Fred Rankin in the surgeon general's office.[22]

At some point, however, the plan took on an additional feature that complicated the already difficult ethical landscape: the journalists were to be paid for their effort rather than working as volunteers. On April 20, 1944, Laurence submitted an application to work for the federal government, and six days later his appointment was approved at a rate of twenty-five dollars per day for no more than 180 days per year.[23] The next day, Laurence swore to "support and defend the constitution of the United States against all enemies, foreign and domestic." He would be reappointed multiple times, until he was inactivated on June 30, 1946.[24]

The surgeon general's office in April 1944 announced the arrangement, through which five science writers were named as "civilian consultants." Although journalists today would look askance at reporters serving as paid consultants to the government, at the time *Editor and Publisher* reported the arrangement without comment or criticism (unsurprisingly, in light of the fact that the magazine's editor until recently had himself served as an NRC consultant), and United Press ran a story that blandly reported that "the writers will work with the surgeon general in the preparation of information for the public on new medical developments in the army." An AP story similarly reported that the five would "work with the Surgeon-General in preparing information for the public on new developments in army medical research."[25]

Potter, who became a member of the consulting group in March 1944, described the setup in an early history of the National Association of Science Writers, writing, "It was a close-knit group who turned out top quality stories on all advances of military medicine; the first uses of penicillin in the armed forces, the miracles of plastic surgery for the severe burn cases of pilots trapped in flaming plane crashes and a hundred other advances that, in the pressure of a full blown war, came to fruition in a few short months and years and represented progress that otherwise might have taken ten and more years to achieve."[26]

In 1944 Laurence racked up eighteen days for the surgeon general, earning $450.[27] "It was nice work, as they say, for while we were civilian consultants (at $25 per diem) when we travelled for SGO we went with a simulated rank of Liet. [*sic*] Col.," Potter later recalled. A simulated rank is status conferred to a civilian equal to that of a military officer—for example, for priority on securing an airplane seat.[28] The NRC operation continued until 1946, although some individual journalists remained consultants to the surgeon general as late as 1954.[29]

By today's standards, the surgeon general's arrangement was highly unethical. But Laurence and other reporters were naturally influenced by the climate of total war in which they operated. They, like their editors and readers, believed that Nazi Germany was an absolute evil whose destruction was an unquestionable good. For Laurence, the issue was personal. Born Lieb Siew on March 7, 1888, he had fled his native Lithuania as a teenager without his family, losing touch with them, and he believed that they had been rounded up by the Nazis. Once he arrived in the United States in 1905, Laurence like many other immigrants had renamed himself, choosing the name William Leonard Laurence to honor William Shakespeare, Leonardo da Vinci, and Lawrence Avenue, the road where he was living at the time. In 1942 Laurence had registered for the draft. (The local draft board registrar noted Laurence's nose, which had been broken by a Cossack when Laurence as a youth had fought in the Russian Revolution of 1905, to "aid in identification.")[30]

Moreover, in the early twentieth century, journalism as a whole was only beginning to grapple with the issue of how interest groups and public relations efforts shaped news coverage and therefore public opinion. Still, some journalists understood that they needed to take steps to ensure that they served the public rather than the government or interest groups. An Oregon journalism professor created a code of journalism that was endorsed by Oregon newspapers in 1922. It said in part, "We believe that the public has confidence in the printed word of journalism in proportion as it is able to believe in the competency of journalists and have trust in their motives. Lack of trust in our motives may arise from the suspicion that we shape our writings to suit non-social interests, or that we open our columns to propaganda, or both. . . . We will resist outside control in every phase of our practice, believing that the best interests of society require intellectual freedom in journalism."[31] Four years later, the Canons of Journalism developed by the American Society of Newspaper Editors and endorsed by the Society of Professional Journalists emphasized that journalists needed to safeguard their independence. "Freedom from all obligations except that of fidelity to the public interest is vital," the document said. The canons also called for journalists to act impartially, asserting, "News reports should be free from bias or opinion of any kind."[32]

An examination of the press releases and news reports written by the surgeon general consultants suggests that the consulting influenced their reporting. Some of the press releases written by Dietz survive, and there is no reason to believe that his were substantially different than those written by Laurence or the other consultant-journalists. Some of Dietz's releases were virtually guaranteed to generate news stories in the general press, such as a release that described an American soldier who had fought for months even though an undetected shell fragment was impinging on his heart.[33] That release inspired coverage by both the Associated Press and United Press.[34] A Dietz release describing an increase in fractures of the bones of the legs or feet due to marching also triggered coverage.[35] As sometimes happens, news coverage did not always strictly follow a press release's message, such as with a press release written by Dietz that played down the risk that soldiers returning to the United States could bring malaria with them and create an epidemic in the homeland. An Associated Press reporter duly reported that fact, but his lead focused on the army's conclusion that the new drug atabrine was better than quinine in treating malaria; malaria was a hot topic among readers.[36]

Other obviously newsworthy releases written by Dietz reported that the number of leg and arm wounds was less in World War II than in World War I or the Civil War, described a new antibacterial treatment for hospitals, and explored development of a new method of freezing whole milk so it could be stocked on hospital ships.[37] Some of Dietz's other releases seemed more designed to appeal to bureaucrats than journalists, such as a press release touting the accomplishments of the Veterinary Division of the Office of the Surgeon General and another about the army's Epidemiology Board.[38]

The army influenced Laurence's coverage of penicillin, which had been discovered in 1928 but received relatively little use in clinical settings. In 1940 a team of Oxford researchers led by Howard Florey published an article in the *Lancet* describing their preliminary work with penicillin.[39] Martin Henry Dawson, a researcher at Columbia, followed up on the *Lancet* article by producing a small amount of penicillin and testing it in two patients. In May 1941 Dawson described the results in a paper presented to the American Society for Clinical Investigation in Atlantic City. Laurence was listening in the audience. "I sat up in my chair when I heard what this strange substance did in the first clinical trials against all kinds of bacteria. It sounded like a nostrum; it sounded like a miracle," Laurence recalled eighteen years later.[40]

But Laurence was not the first to publish the news. He was beaten by Steven M. Spencer, in the *Philadelphia Evening Bulletin*, just hours after Dawson's presentation. Spencer wrote: "From a common mold which grows on stale bread and gives Roquefort cheese its highly prized green color scientists have prepared a germ-chasing medicine."[41] The next morning, Laurence followed with his own story, buried inside the *Times* on page 23, in which he called penicillin "the most powerful non-toxic germ-killer so far discovered."[42]

Government officials in the United States and Britain sought to restrict publication of information about research into penicillin and other medical topics that could be of use to the enemy, including aviation medicine, chemical warfare, and tropical diseases. In 1942 the Office of Scientific Research and Development temporarily suppressed a paper on penicillin submitted to *Science*.[43] Laurence and other reporters played along with the government's desire to avoid mentions of penicillin. Laurence's byline would not appear over a story mentioning the word "penicillin" for more than two years after his breathless discovery story, until August 1943, when he wrote about plans to expand production of the drug. Laurence kept secrets. "The exact amount being extracted from the cheese mold is a military secret, but it certainly is not going to give any comfort to our enemies to state that enough will be made to save thousands of lives among our wounded and fighting men," Laurence wrote.[44]

Overall, the consulting arrangement was a boon to the surgeon general. "By this technique General Kirk had direct pipelines into the AP, Scripps Howard, the Washington Star, the New York Times, The Chicago Daily News, Hearst, and NANA [the North American Newspaper Alliance, a newspaper syndicate], and while the material we wrote went to all papers and wire services as we wrote it, it is understandable that military medicine got major play in our publications too," Potter later wrote.[45] From an ethical perspective, a problematic point is that the arrangement granted Kirk's office silent control over what those journalists were reporting about military medicine. A case in point: in March 1945 Laurence, Dietz, and Potter attended a meeting that had the sole agenda item of reviewing a press release touting the experimental use of antimalarial drugs in inmates at three US prisons. The six-page, single-spaced document seemed unlikely to attract press attention; it was heavy on bureaucratic detail, omitted key information such as the names of the drugs being tested and the success rates of the tests, and skirted discussion of the ethical validity of the tests. Moreover, the fact that inmates were test subjects for antimalarial drugs already had been announced by the Bureau of Prisons the previous year.[46] The surgeon general's release was slated for distribution to both the general press and scientific publications on March 5.[47]

For the *Times*, Laurence covered the news on the mandated release date with a front-page story that said nothing about his role in developing the press release that was the basis for his story. Laurence cited no outside sources, and he did not explore the significant ethical issues about this episode in experiments on prisoners that would be aired more extensively in later decades.[48] Laurence was not the only journalist to gloss over the ethical issues about the prison project. The United Press distributed a story stating that the two hundred convicts had volunteered for the tests. *Life* magazine ran pictures of inmates at Statesville being bitten by infected mosquitoes and ill inmates wracked with chills or high fevers. The researchers "have found prison life ideal for controlled laboratory work with humans," the magazine explained. (During the Nuremberg war crimes trials of Nazi medical researchers in 1946–47, the Nazi researchers' attorney cited the *Life* article to argue that Nazi medical experiments were ethically no more offensive than the US Army experiment that was so highly praised in *Life*.)[49]

In the decades since the malaria experiments, ethicists have debated the experiments. Consistent with the Nuremberg Code, a set of research principles developed after the war, many ethicists have maintained that prisoners are by definition incapable of freely consenting to participate in such an experiment; an apparent minority of ethicists has suggested that inmates can provide such consent. Laurence and the other journalists were not only consumed with supporting the government; they had actively helped prepare the government's public relations messaging on the project. In such a context, how could the journalists be expected to take an independent stance when reporting on that message for their own outlets?

Laurence conceived of himself as a science communicator, not a journalist. Even in the 1940s, many journalists would have understood their job as monitoring and critiquing society and government. But Laurence instead saw himself as part of the process by which research results were transmitted to action—what communication scholars would today refer to as "diffusion of innovation." He saw his mission as giving publicity to new research findings so that they would not be derailed. In fact, Laurence felt that science writers like him had a responsibility to pave the way for social acceptance of scientific discoveries and innovations. In a 1958 speech accepting an award from the American Chemical Society, Laurence maintained that "science writers have an obligation to foresee the social consequences of scientific discoveries—and to prepare the people for them. Had the sociological implications of the industrial revolution been understood, and had the public been alerted to the unemployment problems industrial machines would cause, mankind might have been spared much misery." The journalist must go beyond interpreting science for lay people to promoting its acceptance by lay people, Laurence felt; in other words, the science journalist had a responsibility to promote science, not simply explain it. Laurence told the audience at the award ceremony that the journalist must "bring science to the people, for in a democracy it is the people from which science springs and to whom it returns."[50]

Laurence also held that science journalists had a responsibility to prevent "scientific larceny," or misapplication of credit for scientific discoveries. As an example, Laurence argued that physicists had erroneously been given credit for the development of nuclear energy even though chemists such as Marie Curie had done some of the important research.[51] "I also realized early in my career that I could serve as a mediator between philanthropists and foundations on the one hand and scientists in need of funds on the other. Many a time, when writing a report about a scientific project I would call attention to the fact that financial support was essential to continue the work, and many a project gained such support as a result."[52]

In short, Laurence accepted scientific and technological advancement as an absolute good and saw his proper role as promoting that advancement. Any means to that end was acceptable. Keeping an arm's-length relationship with sources, including the government, was of no concern to him. Here the brilliant Laurence and his editors lost their way. The responsibility of Laurence as a journalist and the *Times* as a news organization was to serve as an independent observer of events. Journalism's mandate is to serve the public, not government. By linking themselves to the government's priorities in supporting the army surgeon general's efforts at shoring up public morale, Laurence and the *Times* failed to serve the public. Laurence and his editors would repeat this fundamental ethical error when the Manhattan Project soon came calling for their help.

3

Magnetic Current

On the morning of January 16, 1944, the Sunday edition of the *New York Times* offered readers a surprise. As was typical at the time, the front page was dominated by war news: British bombers striking German aircraft factories, Germans and Japanese plotting a coup in Peru, the Allies plotting to partition Germany. But stretched across the top of the front page's second and third columns was a very different story, an exciting and upbeat article by Laurence. He breathlessly reported that the Austrian physicist Felix Ehrenhaft had made a dramatic discovery in physics that promised an exciting new energy source: magnetism flows like a current, just as electricity does. Ehrenhaft's finding could be "one of the greatest revolutions in modern science" and could trigger "a new era of technology" such as new types of machines powered by magnetic current, Laurence pronounced.

With the story, Laurence kicked off a months-long campaign to draw attention to Ehrenhaft's work even though mainstream physicists already had rejected the research as flawed. Laurence's stories would trigger waves

"All the News That's Fit to Print."

The New York Times.

LATE CITY EDITION

Section 1

VOL. XCIII...No. 31,483. NEW YORK, SUNDAY, JANUARY 16, 1944.

WALLACE TELLS CIO VICTORY AIM ALONE GUIDES PRESIDENT

Statement at Parley Here Is Construed as Apology for Labor-Draft Proposal

STIRRING DISCORD SCORED

Vice President Attacks 'Some' Business Men, Lauds Others —Farm Lobby Disowned

Proof Offered of Existence Of Pure Magnetic Current

Discovery by Prof. Felix Ehrenhaft, if Confirmed, Expected to Rank With Faraday Finding of Dynamo Principle

By WILLIAM L. LAURENCE

STETTINIUS ASKS FULL COOPERATION WITH SOVIET RUSSIA

In Broadcast He Calls Any Other Course in War or Peace 'Tragic Blundering'

BIG DEPARTMENT SHIFTS

For-Reaching Plans of Reorganization Are Announced by Secretary Hull

RAF BOMBERS SMASH BRUNSWICK, DROP 2,240 TONS IN 23 MINUTES; FRENCH IN 5TH ARMY TAKE TOWN

NEW SOVIET DRIVE

Foe Reports a Russian Push in the Leningrad-Lake Ilmen Sector

PRIPET GAINS CONTINUE

Heaviest Fighting Still Taking Place on Vinnitsa Front as Red Army Holds Firm

Acquafondata and 3 Peaks Captured in Advance in Italy

French Troops Meet Fiercest Resistance on San Pietro Ridge as Americans Battle for Mt. Trocchio—Adriatic Area Quiet

700 PLANES STRIKE

British Deliver Record Concentration at Reich Industrial Center

MOSQUITOS HIT AT BERLIN

Magdeburg Also Attacked in Double Feint—French Coast Battered Night and Day

MORE COAL COMING TO THE NORTHEAST

Ickes Also Requests West Virginia Producers to Hold Remainder for Emergencies

DEMOCRATS MOVE TO KILL P.R. VOTING

City Organizations Inspire a Move at Albany Along Two Legislative Lines

Peru Charges a Revolt Plot By Germans and Japanese

AUSTRALIANS DRIVE TO SIO'S OUTSKIRTS

Seize Much Booty in Push to Last Japanese Base on Huon Coast in New Guinea

War News Summarized

SUNDAY, JANUARY 16, 1944

Unseasonal Thunder Causes 'Blast' Alarm

Senate Puts 'Field Amendment' In Tax Bill, Cutting Loss-Taking

3-Power Occupation of Germany Reported Considered at Teheran

Figure 3.1. Laurence's front-page article on the supposed discovery of "magnetic current," which ran on the upper left half of the front page for January 16, 1944, kicked off a years-long effort by him to focus scientific attention on a line of research that eventually proved fruitless. (From The New York Times. © 1944 The New York Times Company. All rights reserved. Used under license.)

of media coverage by other journalists in the United States and abroad over the next months and years—but the research would turn out to be a dead end, as evidenced by the fact that today we do not use exotic devices powered by magnetic current.

The scientific establishment regarded Ehrenhaft warily. In the early 1900s, while a faculty member at the University of Vienna, he had engaged in a pitched academic battle with the physicist Robert A. Millikan, then located at the University of Chicago, over the nature of the electron. Physicists gradually coalesced around Millikan's findings. The dispute received little news coverage, likely in part because few journalists then focused on reporting about science.[1]

The extended argument left Ehrenhaft with a poor reputation among many physicists, who felt that he had been contentious and closed-minded. One with a low opinion of Ehrenhaft was Warren Weaver, a mathematician who at the time directed the Division of Natural Sciences at the Rockefeller Foundation. In the 1930s the foundation had awarded a grant to Ehrenhaft. After Ehrenhaft immigrated to the United States, in 1939 a researcher at the University of Iowa wrote Weaver to ask that the foundation consider supporting him. But Weaver replied that he did not "consider it very likely that [Ehrenhaft] would accomplish anything of particular importance during the remainder of his life." Weaver also stated that many physicists had viewed Ehrenhaft's long dispute with Millikan as "a stubborn and ill-advised controversy."[2]

By this point, Ehrenhaft's research had moved away from electrons to magnetism. He took the unconventional view that magnetic poles could exist in isolation from one another, in a form known as a monopole, and that these monopoles could flow in a current somewhat akin to electricity. Ehrenhaft made repeated presentations at physics conferences in which he demonstrated phenomena that he interpreted as contradicting the traditional views of magnetism. Some of these presentations—such as a June 1941 talk at Brown University and a May 1942 presentation at Johns Hopkins University—escaped media attention. Others did receive coverage: the *Times* and Science Service, a nonprofit news service about science, briefly covered a talk by Ehrenhaft at Columbia University in 1940 in which he reported that light had moved particles of matter, which he characterized as a manifestation of his view of the nature of magnetism.[3]

The theoretical physicist P. A. M. Dirac, who also had proposed the existence of magnetic monopoles but had little interest in Ehrenhaft's work, recalled that Ehrenhaft got little traction with his presentations. "His reputation had sunk so low, everyone believed him to be just a crank. All he could do was to buttonhole people in the corridors and pour out his woes. . . . He kept saying that he had these experimental results and nobody would listen to him."[4]

In March 1941 Ehrenhaft published a short paper in *Nature*—a general science journal, not one specializing in physics—reporting that he had used ultraviolet light to magnetize small metallic objects such as paper clips. The *Times* ran an unbylined, one-paragraph report, which Ehrenhaft's wife cited to request more financial support for Ehrenhaft from the Rockefeller Foundation. The skeptical Warren Weaver still held sway there, and the foundation said no.[5]

But the article grabbed the attention of Laurence, who took pride in combing journals and attending conferences in search of research that

Figure 3.2. Laurence in his *Times* office, in an undated photo. (New York Public Library)

other journalists might miss. "The article was very interesting and appeared on the face of it to be a highly important and revolutionary development in science, of incalculable practical importance, *if true*," Laurence later recalled. But Laurence "could not get the enthusiastic comment I expected and I gathered at the time that Dr. Ehrenhaft's report should be taken with a large grain of salt," so Laurence did not write an article.[6] Laurence deemed a second paper by Ehrenhaft "even more startling" but "decided to sleep on it until Dr. Ehrenhaft could present more convincing proof."[7]

When Ehrenhaft spoke at a meeting of the American Physical Society at Columbia University in New York in January 1943, Laurence attended with great interest. "It was the sort of material that very seldom falls in the lap of a reporter, a discovery that, if true, would mark one of the greatest landmarks in human progress," Laurence later wrote. But once again Laurence could find no scientist willing to vouch for the validity of Ehrenhaft's research, and so once again he decided not to report on it.[8]

Months later, Laurence noted that Ehrenhaft was again scheduled to speak, this time at the January 1944 meeting of the American Physical Society. "At this point it occurred to me that I had not been quite fair to Dr. Ehrenhaft and that he was entitled to the same treatment I would give to any scientist of his standing in the United States," Laurence later recalled. The reporter contacted the scientist, who welcomed him to his laboratory in New York, where he got to look through the apparatus and even control it.[9]

Laurence impressed Lilly Rona, Ehrenhaft's wife and tireless advocate. In her journal she wrote of Laurence: "A little man, reminding of Michelangelo, with his smashed nose, listened with highest concentration to the vivid explanations of Ehrenhaft." When Laurence asked about objections to the research that had been raised by the eminent physicist Francis F. G. Swann, Rona complained that Swann had not bothered to view the experiments himself—an objection that Laurence found telling, because later he would repeat it when justifying his decision to publish an article on Ehrenhaft. Before leaving Ehrenhaft and Rona, Laurence promised "to write a series of articles" about the magnetic current research, a promise that left Rona ambivalent. She wrote in her diary, "One does not know the powers working backstage—but I had still the impression that if he could do as he wants—he would write an intelligent article about our work."[10]

Five days later, Laurence attended the scientist's talk, which Laurence described as "a sad and painful spectacle" because two-thirds of the audience exited before Ehrenhaft spoke. Laurence interpreted the exodus as evidence of closed-mindedness or perhaps even anti-Semitism. After the talk, Laurence interviewed physicists, who admitted that none had tried to replicate his work—a finding that reinforced Laurence's impression of physicists being closed-minded.[11]

As a result, Laurence wrote the blaring story that greeted readers on the morning of January 11, 1944. The article was highly visible: it occupied two columns on the top of the front page, positioned between articles on war policy and labor politics—an unusual location for a science story on any day of the week, much less the Sunday edition. "Proof offered of existence of pure magnetic current," trumpeted the headline. The article described a series of demonstrations by Ehrenhaft to the assembled physicists that, Laurence wrote, showed the existence of magnetic current. For example, a magnetized needle floating in an acidic fluid generated bubbles of hydrogen and oxygen, which Ehrenhaft interpreted as evidence that magnetism flowed through the needle and caused water to decompose into its chemical components. This demonstration "greatly interested the physicists," Laurence wrote.[12]

Laurence's lead did contain some journalistic detachment; Laurence himself did not flatly claim that Ehrenhaft's work was revolutionary but instead put that claim into Ehrenhaft's mouth. Specifically, Laurence cited Ehrenhaft's experiments, "which he said provided for the first time experimental proof of the existence of pure magnetic current." Laurence later would invoke his use of attribution, such as the phrase "he said" in the quote in the previous sentence, as evidence that he was acting as a neutral reporter and was not himself arguing for the validity of Ehrenhaft's conclusions. But that subtle note of journalistic detachment was overwhelmed by the overall tone and approach of Laurence's story, which clearly favored Ehrenhaft's conclusions. The article quoted no critics of Ehrenhaft's work; indeed, the article quoted no one by name other than Ehrenhaft. And it depicted the physics community as reacting to Ehrenhaft's demonstrations with wide-eyed wonder: Ehrenhaft "created a sensation among the prominent physicists" who witnessed his presentation, Laurence wrote. "They said that if the experiments described by Professor Ehrenhaft could be corroborated by others they would mark one of the greatest revolutions in modern science, to be ranked with the discovery of the principle of the dynamo by Michael Faraday 113 years ago." Ehrenhaft's work "would mean the ushering in of a new era in technology."[13] The *Times* wasn't the only news outlet to cover Ehrenhaft's paper. The Associated Press's Howard Blakeslee also prepared a similarly breathless account, but the AP article got much less play. The *Los Angeles Times*, for example, ran it on page 13—a big difference in emphasis from the New York paper's front-page placement.[14]

"What a sensation Front Page Article!" Rona exclaimed in uneven English in her diary. Ehrenhaft "came to my bedroom and I read the whole article to him. We were thrilled—So wonderfully written—So clear—And on a so important place!"[15] Her diary makes no mention of Blakeslee's coverage.

The coverage by the *Times* and AP spawned newspaper and magazine stories around the nation. Before long, magazines—with their longer lead time—began to catch up on the story. Citing Laurence's story, the *New Yorker* declared Ehrenhaft to be "quite possibly the man of the century."[16] *Harper's Bazaar* ran a photograph of Ehrenhaft peering into a laboratory instrument. "If corroborated, the discovery will mark one of the greatest scientific revolutions since Faraday's discovery of the principle of the dynamo," the magazine stated. *Time* took a bit more measured approach, stating that Ehrenhaft "claimed to have proved that magnetism moves."[17] The *Times* article inspired the magazine *Liberal Judaism* to run a lengthy profile of Ehrenhaft in its February 1944 issue. Comparing Ehrenhaft to Einstein and highlighting his Jewish heritage, the article stated that "no one during the past three decades has been able to refute the correctness of his findings."[18]

Laurence drew static from scientists over his article. Rona wrote in her diary that Laurence told her "he had received very many letters, but from unimportant people" complaining about his article." Laurence told her that he had told one of the complainants, an unnamed Dartmouth College researcher, that he "seems to suffer from scientific arterial sclerosis."[19] The researcher likely was Dartmouth physicist Gordon Ferrie Hull, who later told a fellow physicist that his misgivings over Laurence's coverage of Ehrenhaft and of an unrelated story about Einstein had inspired him to write to the *Times*. The letter "was not published but was turned over to Laurence, who thereafter wrote me a very abrasive letter in which he indicated that compared with Ehrenhaft I was a very low grade physicist. . . . There is no question but that he has a strong bias in favor of Jewish physicists."[20]

Still in need of financial support, Ehrenhaft used Laurence's coverage as a justification for inviting the president of the Rockefeller Foundation in February 1944 to visit his laboratory. Weaver declined on behalf of the president, saying that the foundation could not consider "requests which fall outside our present program."[21]

In February, Laurence's colleague at the *Times*, Waldemar Kaempffert, weighed in on Ehrenhaft, describing research published in *Nature* by a British researcher who "confirms Ehrenhaft's observations . . . but differs completely with the interpretation." The British researcher, James T. Kendall, had concluded that the bubble movement that so excited Ehrenhaft was caused by currents in the liquid surrounding the magnet, not because of the magnetic current that Ehrenhaft thought existed. Kaempffert took no stand on which interpretation was correct, and the article ran deep inside the paper.[22] In her diary Rona blasted Kaempffert's article as containing "the most obvious misconceptions" about Ehrenhaft's work. Kendall, she wrote, was so focused on "his high mission and ardent desire to refute the Antichrist—the dangerous Ehrenhaft—that he lost his head—with all its inventory." Ehrenhaft, on the other hand, simply laughed off the criticism, she wrote.[23]

The next month, Ehrenhaft delivered yet another presentation, this time to the American Institute of Electrical Engineers. He criticized physicists, holding that their view of magnetism was little more than a "fairy tale." A *Times* article on the talk—unbylined but presumably by Laurence—was placed on page 7, with a headline guaranteed to irritate physicists: "Old Rule in Physics Called 'Fairy Tale.'"[24]

Weaver had been simmering with anger over the fawning press coverage, but matters came to a head when a reporter for the Scripps-Howard newspaper chain wrote a laudatory feature on Ehrenhaft and his wife. The article did not examine the validity of Ehrenhaft's scientific claims but depicted him as battling against entrenched physicists who, "with scarcely an exception, turned deaf ears to the distinguished Austrian scientist." The article described scientists walking out on Ehrenhaft's January 1944 talk at Columbia, implying that they were unwilling even to consider his evidence. His main advocate was his wife, who was not even a scientist, the article said. Even more inflammatory for physicists were the headlines that ran over the story in various newspapers. The *New York World-Telegram*, for example, ran the story on March 16 under the headline "Wife Helps Scientist Prove 700-Year Belief Is Wrong on Magnetic Current." Two weeks later, the *Pittsburgh Press* ran the article under the similar headline, "Wife Helps Scientist Prove Belief Wrong on Magnetic Current."[25]

Weaver, still on the staff of the Rockefeller Foundation, vented to Henry A. Barton, the director of the American Institute of Physics. "When the Times (of all papers) put Ehrenhaft on the front page of the Sunday edition a couple of months ago, I was shocked," Weaver wrote. "When the World Telegram puts this drool on the front page of their second section, with its indecent use of the good honest word 'prove,' I am nauseated." Weaver implored Barton to reach out to newspapers to "please not to make such a cheap travesty of science; and incidentally not to make such asses of themselves."[26] Barton noted that the strong reaction by physicists might have been counterproductive. "I am afraid that such an attack has been made by physicists on Lawrence [*sic*] that even the editors of The New York Times resent it." Barton said that "Lawrence has a complex on the possibility that someone is being persecuted. I did my best for a half hour to convince him that Ehrenhaft's stuff is bunk," but "I do not think anyone could have persuaded Lawrence out of writing it up." Barton also warned that criticisms of Ehrenhaft could come back to haunt the physicists. "We cannot afford to answer by casting doubt on Ehrenhaft's honesty or sanity because of the very real danger that we might become involved in a lawsuit," he wrote.[27]

Weaver then approached Arthur Hays Sulzberger, the publisher of the *Times* who also served as a trustee of the Rockefeller Foundation, to complain about Laurence's coverage. Sulzberger asked Laurence to meet with Weaver, and the two met for two hours on April 13, 1944, which Laurence described in a six-page memorandum to Sulzberger that related the history of Laurence's coverage of Ehrenhaft from 1941 as well as Laurence's meeting with Weaver earlier that day. Laurence asserted that he and Weaver had agreed that the article was "100 percent accurate. The main point on which difference of opinion does exist is whether we should have ignored the story altogether or at least played it down," because the experiments "contradict too many widely accepted views about magnetism."[28] Unsurprisingly, Weaver had his own take on the meeting with Laurence. In his first draft of a memorandum to Sulzberger, dated April 17, Weaver termed the Ehrenhaft article "a quite unnecessary mistake" because many physicists could have warned Laurence off the story; "a spectacular mistake" because of the story's placement on the front page of the Sunday edition; and "a dangerous mistake" because it could mislead the public "about science and scientists." Weaver did not disagree that Laurence had reported Ehrenhaft's statements correctly, but he argued that the article nevertheless had provided a misimpression: "When the Times puts a science story on the front page of the Sunday edition, the average reader concludes that it is important." The fact that *Science* had published Ehrenhaft's research was no defense for covering it, Weaver wrote, adding that "the editorial policy of Science is well known to have been at once liberal, undiscriminating and erratic." Weaver also rebutted Laurence's suspicion that the opposition to Ehrenhaft was rooted in his status as a refugee. "A great many European refugee scientists have come to the U.S. over the last ten years, including a good many German scientists of Jewish blood. I think we can be very proud of the generous, friendly and open-minded way they have been received." Weaver also questioned the significance of the demonstrations that Ehrenhaft had conducted for Laurence. "To 'see with one's own eyes' is impressive but dangerous, especially when the phenomena are subtle and complicated. . . . Magnetic Theory is a complicated and difficult subject; and significant answers to these questions can be furnished only by experts of proven ability and reliability."[29]

Before sending his memorandum to Sulzberger, Weaver shared the draft with two physicists in New York City. One cannot be identified from the archival record, but the other was George Pegram, a prominent physics professor at Columbia University, who endorsed Weaver's arguments, saying his only revision would be to emphasize that physicists had been open-minded toward scientific refugees. "It would be a thankless job, but I should be glad if some competent person or persons would look into the experiments by Ehrenhaft and point out rational explanations for them," Pegram wrote. "Had physicists not been so busy on account of the war, I venture to say that someone would have done this before now."[30] Pegram was no fan of Laurence; four years earlier, Pegram had complained about Laurence's reporting on atomic energy research. After receiving Pegram's response, Weaver revised his memorandum to ask that his involvement be kept confidential, and he said that his letter had to "terminate my connection with the affair."

While this correspondence was unfolding, the two science reporters at the *Times* were still dueling over the story. On April 23 Kaempffert revisited the issue, apparently not due to any new news on the topic. Asserting that the *Times* science department "takes no position in the controversy," Kaempffert nevertheless concluded that "probably the physicists are right." He criticized the physics establishment for arguing against Ehrenhaft "in terms of dogma" rather than performing experiments that could definitively prove or disprove his claims.[31] Six days later, on April 29, another *Times* story—unbylined but presumably by Laurence—covered a presentation by Ehrenhaft at a meeting of the American Physical Society in Rochester, New York, where the physicist reported that he had used a magnet to decompose water into bubbles of hydrogen and oxygen. More oxygen accumulated at the magnet's north pole than at the south pole, which he said was evidence of a magnetic current moving through the water from one pole to the other. The article quoted objections from a physicist who said that he tried and failed four times to replicate Ehrenhaft's results. The story ended with Ehrenhaft responding that he had not yet published full details of his experiments so no one would have been able to reproduce them.[32]

Although Weaver had written in his April 24 letter that he was ending his involvement in the controversy, on May 1 Sulzberger wrote him to express his perplexity about the Rochester conference, "where Dr. Ehrenhaft seems to have been the chief attraction. If it is 'no dice,' why is he afforded such an important platform twice in a relatively short time? It certainly suggests that there is news interest in what he does."[33] Weaver offered two explanations: "a) that Dr. Ehrenhaft is a very insistent individual and, b) that (contrary to Mr. Laurence's present unhappy conviction) American physicists are exceedingly hospitable, open-minded, and patient."[34]

Simultaneously, Ehrenhaft's wife again tried to leverage Laurence's coverage to obtain funds from the Rockefeller Foundation. In a letter to the foundation's president, she complained, "After Dr. Ehrenhaft has pronounced the result of his work of more than thirty five years, now that the importance of the discovery of the Magnetic Current has been stated by men of highest experience here and abroad, published all over the world and broadcasted from London in 25 languages with an exact description of the facts . . . now you reject his request for your assistance for the third time." In an internal memorandum at the Rockefeller Foundation,

Weaver held firm against providing support: "I can only say that I do not know a single reputable physicist in the United States to whom I could turn with any hope that he would sponsor a request to support Ehrenhaft." On April 19 the foundation president wrote Mrs. Ehrenhaft that the foundation had devoted "careful and extended consideration" to the question but had denied the request. "The negative decision is final," the president wrote, obviously hoping to put the long-festering issue to rest.[35]

Despite Weaver's intervention, media attention on Ehrenhaft continued. The findings even touched science fiction fans. Their leading magazine, *Astounding Science-Fiction*, ran a lengthy article in its May 1944 issue explaining Ehrenhaft's work.[36] In June 1944 the monthly magazine *Popular Science* jumped on Laurence's bandwagon with a story about Ehrenhaft titled "Magic with Magnetism." The article related the author's eyewitness account of what he saw when watching Ehrenhaft's experiments; the article is thoroughly illustrated with photographs of Ehrenhaft at work and with close-up photos of his laboratory apparatus. The article quoted no critics or skeptics.[37] That same month, the magazine *Illustrated* ran two pages about Ehrenhaft's work, accompanied by five photos of the scientist at work in his laboratory and gushing about the technology's promise. "When the magnets of your car begin to run down, you will pull up at a recharging station, couple up, switch on, and recharge your magnets in a few seconds."[38]

Meanwhile, Laurence again took to the *Times*'s news columns to tout Ehrenhaft's work. On June 25 he reported a presentation by Ehrenhaft to the American Physical Society in Rochester, New York, in which the physicist caused a microscopic drop of liquid chemical containing iron to continually move between the two poles of a magnet. The test showed the feasibility of a motor running on magnetism rather than electricity, the reporter asserted, adding a jab at physicists that the demonstration was "so simple that a high school student of physics could repeat it without difficulty."[39] A month later, on July 30, Laurence reprised the evidence that Ehrenhaft had presented in Rochester, calling it "the most dramatic evidence so far in support of his hypothesis." Laurence also insulted the conventional physicists who remained skeptical of Ehrenhaft. "These observations, all based on hundreds of experiments, present a formidable array of evidence that can no longer be dismissed in the manner of the incredulous farmer who, on seeing a giraffe for the first time, exclaimed:

'There ain't no sich [*sic*] animal!' . . . It is possible that other explanations may be found for the observed phenomena. But there can no longer be any question about the accuracy of the observations."[40] Swann, whose imprimatur could be particularly valuable, wrote to Laurence to request a copy of the article. Laurence sent the story, noting his "hope to have the privilege one of these days to discuss the matter with you in confidence."[41]

Ehrenhaft was ecstatic about the article. "It is a psychological masterpiece in itself," his wife wrote Laurence. "We enjoyed reading it over and over again and laught [*sic*] heartily about the parabel [*sic*] of the giraffe which fits the occasion so perfectly." But then she proceeded to exhibit her own biases. "Your article came just in the right moment to trouble those who fight us under the flag of the Am. Phys. Society and its Life Guards the Phys. Review—Millican's S.S."[42]

That excitement was short lived. By September, Ehrenhaft grew despondent about the difficulty in publishing his results, and he turned to Laurence for advice. "Laurence received us very kindly and told us to be patient . . . patient . . . patient. We should not think that the greatest revolution in science since 1700 years could be won over night," Ehrenhaft's spouse wrote in her journal. Laurence said that Swann was moving toward supporting their research and then made the remarkable claim that Albert Einstein was interested in Ehrenhaft's work. "I am of the opinion that Einstein cannot go ahead in his work without your new knowledge," Laurence told Rona. For her part, Rona suspected that Einstein was simply jealous of the attention the *Times* was giving Ehrenhaft.[43] In any case, it is unlikely that Einstein placed great stock in Ehrenhaft's work; correspondence between the two indicates that Einstein was deeply skeptical of Ehrenhaft's claims.[44]

In January 1945 Laurence covered a talk by the researcher to a meeting of the American Physical Society at Columbia University. Ehrenhaft reported that light beams could cause microscopic specks of matter to move, a finding that Ehrenhaft said supported his theory of magnetic current. Someone at the Rockefeller Foundation—presumably Weaver—annotated a copy of Laurence's article with the trenchant comment, "Bill Laurence is surely doing his best to promote Ehrenhaft into a public success."[45] But the AP's Blakeslee also covered Ehrenhaft's claims, reporting that Ehrenhaft was advancing "a new concept of the motion of light." Blakeslee's article appeared on inside pages in the *Washington Post*, the *Baltimore*

Sun, and the *Los Angeles Times*. A few days later *Time* also reported the paper, but with a caution: "Pondering this revolutionary theory, and well aware that Dr. Ehrenhaft is a cantankerous man in an argument, his fellow physicists kept skeptically mum."[46]

The story continued to diffuse into the technical and hobbyist press. In April 1945 the magazine *Radio News* took up the topic, reporting that "the possibility of magnetic ions and magnetic current is a controversial subject, but there is experimental evidence that they do exist."[47] In April 1945 a brief item in *Popular Science* announced that Ehrenhaft's findings had been "confirmed" by experiments conducted by researchers at Manhattan College and the Federal Communications Commission.[48] In June 1945 the AP's Howard Blakeslee reported on a talk by Ehrenhaft at an APS meeting in Columbus, Ohio, arguing that rays of sunlight carried a magnetic charge. *Newsweek* also covered the talk.[49]

Still in financial straits, Ehrenhaft in 1947 again asked the Rockefeller Foundation for financial support, this time for him to return to postwar Vienna to resume his research there. Weaver declined, telling Ehrenhaft that it was unclear whether the foundation would be able to operate in Austria. "In any event it is altogether unlikely that we would consider a grant of this character in the physical sciences," he wrote. In an internal memo, Weaver wrote that "this man is very aged, has a most unsavory reputation. He has pestered us practically continuously."[50]

After 1945, coverage of Ehrenhaft died down. By this point, Laurence had moved on from covering Ehrenhaft. As we will see, in mid-1945 Laurence was recruited by the US government to write the press releases for the use of atom bombs at the end of World War II, and after the war he made atomic energy the focus of his reporting. He likely had little time or inclination to pursue the Ehrenhaft story.

Ehrenhaft returned to Vienna in 1947 to resume teaching and research at the University of Vienna. US intelligence officials investigated his research there and recommended keeping him at arm's length from US occupation officials while remaining ready to capitalize on Ehrenhaft's research if it panned out.[51] He dropped off the radar screen of American science journalists, although Rona remained in the United States and continued to try to convince journals to publish Ehrenhaft's research. Ehrenhaft died in Vienna in 1952, and Rona died in New York in 1958.[52] After Ehrenhaft's death, the *New York Times* ran an obituary written by the Associated

Press. The six-paragraph item reprised his claims about magnetic current in 1944 but did not note the controversial nature of his research. "His research into the nature of electrical energy and his early work on the electron contributed indirectly to the later discoveries of scientists preparing the groundwork for atomic energy," the article stated.[53]

Laurence's aggressive promotion of Ehrenhaft permanently tainted how some physicists viewed the journalist. In 1958—a decade and a half after Laurence's front-page coverage of Ehrenhaft—physicist Samuel A. Goudsmit chided Laurence about it. Laurence had referred G. A. Zotos, an outside-of-the-mainstream researcher, to Goudsmit, who was famous for his early work on the spin of electrons as well as his wartime work investigating the German nuclear program on behalf of the Manhattan Project. In a letter, Goudsmit labeled Zotos a "crack pot" like "Ehrenhaft, whom you gave so much space in your paper years ago. That man wasn't even harmless!"[54] Another critic of Laurence was Raymond T. Birge, a physicist at the University of California, Berkeley. Birge commiserated with Gordon Ferrie Hull, the Dartmouth physicist whom Laurence had lambasted after he complained about the initial Ehrenhaft story. "Usually the good newspaper scientific writers consult the leading scientists of the country when they want to know whether any matter is authentic or not, and certainly Laurence should have no difficulty in obtaining the true facts about Ehrenhaft from any reputable physicist," Birge wrote two years after the episode.[55]

Laurence and Weaver saw little of one another for some time but finally reconciled in March 1951, when they met at an awards ceremony in Princeton, New Jersey. "Mr. Laurence was completely silent for a short time," Weaver related in his diary. Then Laurence turned to Weaver and said, "The last time you and I had an important conversation, you were 100 per cent right and I was 100 per cent wrong." Then the two chatted amiably at length.[56]

4

ATOMLAND-ON-MARS

In early 1945 readers of the *Times* encountered articles by Laurence on a regular basis. One day in January he covered one of Ehrenhaft's papers on his theory of magnetic current; the next day he reported on a lecture by the physicist I. I. Rabi describing how atomic clocks could be constructed. Laurence rhapsodized that Rabi had provided "blueprints for the most accurate clock in the universe, tuning in on radio frequencies in the hearts of atoms and thus beating in harmony with the 'cosmic pendulum.'" Laurence's coverage of Rabi exhibited "wonderment typical of atomic age discoveries," in one analysis.[1] In March he wrote two stories on cancer research and then one on research into biotin, a career-long interest of his.

But after March 10, 1945, nothing would appear in the *Times* under Laurence's byline for six months. When his byline rematerialized, it would be over a more sensational story than he had ever published before: the atom bomb.

At the time when Laurence's journalism vanished from public view, he had deduced that the government had a classified project on atomic technology under way. As World War II progressed, physicists he had known for years had begun to vanish. "I would come to a meeting and a couple of top scientists who had always been there were missing," he recalled in an oral history interview decades later. "I would call up their universities and laboratories to talk to them, and I would get evasive answers . . . so I began putting two and two together and figuring that there must be something going on."[2]

Then a nephew of a secretary at the *Times*, whom Laurence had helped choose science courses at college, told Laurence that he was leaving for a secret location in Oak Ridge, Tennessee, that was conducting research on uranium. Laurence leveraged that tidbit in conversations with scientists to act as if he knew more. "In that way I got a very good general over-all picture."[3] The reporter found out that planes were being barred from flying over New Mexico, and he learned about the existence of secret facilities in Washington State, Oregon, and California.

Although Laurence did not try to reveal these secrets about the US atom-bomb program in the pages of the *Times*, he did try to publicize what he had found out about the Nazis' atomic-bomb program. However, the federal government's Office of Censorship opposed news coverage of even the enemy's A-bomb efforts out of concern that such stories could risk drawing attention to America's huge but ultrasecret atomic-weapons project. Under the Office of Censorship, news organizations sought government review of news reports that might have posed a potential risk to the war effort; although the media's participation technically was voluntary, the government always had the last word.[4] "I kept writing stories speculating on the secret weapons that Hitler and Goebbels were boasting would soon win the war for them," Laurence wrote later in his book *Men and Atoms*. All were submitted to the Office of Censorship, headed by Byron Price, former executive news editor of the Associated Press. "Story after story I submitted to Byron Price's office was returned with the request not to publish. Even speculation on what the enemy might be doing was out of bounds, since, as was explained to me, we did not want the enemy to know what we knew about him."[5] The Office of Censorship's records show that in December 1943 and August 1944 it rejected stories by Laurence about the potential of using U-235 in explosives. After the second time, *Times* managing editor Edwin James wrote that Laurence would "commit suicide" if the *Times* were scooped on the story.[6]

Other reporters also discovered traces of the Manhattan Project and synthesized those bits of evidence into a larger picture.[7] But given his understanding of atomic physics and his access to top researchers, Laurence was perhaps uniquely positioned to understand those bits of evidentiary string. He leveraged those insights on an unseasonably warm Friday the 13th in April 1945, the day after President Franklin D. Roosevelt died suddenly, elevating Harry S. Truman to the presidency, when Laurence received an offer that changed his professional life.

The offer came from army Lt. Gen. Leslie R. Groves, the head of the Manhattan Project. Although Laurence knew well many of the leading scientists in atomic physics, he had fewer acquaintances among military officers and had never met Groves—so Laurence did not recognize the portly general who strode into the *Times* newsroom that day for an appointment with James. Groves had decided that he needed someone on his staff to write press releases about the first use of the atom bomb. Groves's first choice was Jack Lockhart, who worked in the Office of Censorship, but Lockhart begged off and suggested Laurence instead.[8] "While we had in the project a number of competent men with sound newspaper backgrounds, they already had more than enough to do, and it seemed to us, in any case, that it would be much better to bring in an outside newspaperman who would have a more objective touch," Groves recalled in his 1962 memoir.[9]

Groves also might have been attracted by Laurence's proven track record for dramatic writing; the physicist Norman F. Ramsey later said that Laurence was chosen to make "a deliberate effort to make it [the bomb] more dramatic than it was."[10] Laurence himself offered a typically self-serving interpretation of his selection: "Some wag remarked that since I knew too much about it, General Leslie R. Groves, the commanding general of the atomic bomb project, had the choice of either shooting me or hiring me."[11]

Whatever the reason, Groves asked James—without Laurence in the room—to allow Laurence to work on a classified project for a period of time. James apparently pushed back about the length of time Laurence would be away. After the end of the war, in the happy afterglow of the American victory, Groves joshed to Arthur Hays Sulzberger, publisher of the Times, "Will you please tell Mr. James that I hope he now agrees that this story required more than a few days?"[12]

Groves and James discussed how Laurence would be compensated. Groves wrote in his memoir, "It seemed desirable for security reasons, as well as easier for the employer, to have Laurence continue on the payroll of the *New York Times*, but with his expenses to be covered by the MED [the Manhattan Engineer District, the formal name of the Manhattan Project]."[13] In 1962, when Groves's memoir was published, Laurence would claim that this arrangement was unknown to him—and then ask the *Times* for money that he said the *Times* owed him.

When Groves spoke directly with Laurence about coming to work for him, the newspaperman cheekily insisted, "If you want me to do any writing, I must be given access to first-hand sources. I hope you'll permit me to go to Tennessee, Washington and New Mexico." At that demand, which indicated that Laurence already understood at least the general outlines of the Manhattan Project, "General Groves winced," Laurence wrote.[14] Laurence's demand was clearly a bluff; he never would have turned down the opportunity. Groves did grant Laurence the wide access that he requested, directing Col. Kenneth D. Nichols at Oak Ridge, "It is requested that Mr. Laurence be given access to the various installations in the United States and to all pertinent information and data, except such information or data as would disclose other than general details concerning processes and formulae. There is no objection to Mr. Laurence having discussions with any individual connected with the project."[15] However, Groves did make some topics off limits, such as the Combined Development Trust, an effort by Groves to give the United States control over world supplies of uranium ore; and Alsos, an intelligence project in occupied Europe to determine how close Germany had come to developing an atomic weapon.[16]

Laurence was away from the *Times* newsroom for more than four months. "No one knew where he was and, after a week, I concluded he must be following a story he wanted to keep under wraps. He had left his desk, however, uncharacteristically clean. Others noted his continuing absence, but our curiosity was met by Edwin James with evasions," his friend Arthur Gelb recalled.[17] Laurence's science-writing colleagues at other media organizations, including those still working as consultants to the surgeon general, also noticed.

Laurence's new gig in Washington, DC, started on May 8. While others were celebrating Victory in Europe Day marking the German surrender the preceding day, Laurence reported to Groves's office on the fifth floor of what was then the War Department Building and today is the headquarters of the State Department. There he received an overview briefing on the Manhattan Project.[18] Next he was off to Nashville, Tennessee, where a car brought him to Nichols in Oak Ridge on May 10. Laurence later wrote, "I called on him at his office and then he took me around, and what I saw was simply staggering to the imagination. He took me on a hilltop where I could see the whole city in a valley below."[19] Nichols showed Laurence the mass spectrometers—devices using powerful electromagnets for separating the rare U-235 atoms from the far more common but less active U-238.

George O. Robinson, a public affairs officer for the MED in Oak Ridge, recalled, "Mr. Laurence began his quest for official facts with visits into the huge, restricted Oak Ridge plants. The quest then took him to Los Alamos, Hanford, Berkeley, California; Chicago and New York, and back again to Oak Ridge, where he wrote the majority of his stories in a special office set aside for him in District headquarters."[20]

Laurence later told Gelb that he slept poorly at the start of his Manhattan Project work, worried about how he would perform. Laurence obsessed about the lead—the first paragraph, or first few paragraphs of a news story, which traditionally announces the key news being reported but also helps frame or interpret that news. "I devised dozens of leads, but every one sounded phony. I began to believe that a lead couldn't be written." He suffered nightmares about being in a Western Union office to transmit his story about the atom bomb but being unable to write a lead. "For two weeks I had nightmares in which I was at the typewriter, with the biggest story in the history of the world—and I couldn't think of a lead for the story. But such was the indoctrination in security at Oak Ridge that after two weeks, I stopped dreaming about it."[21]

On May 14, with the Allies concentrating on the defeat of Japan, the federal government's so-called Interim Committee, which had been appointed by Secretary of War Henry L. Stimson to consider atomic-weapons issues until a permanent governance structure could be set up, decided at its second meeting that the upcoming Trinity test—the first detonation of a plutonium bomb—needed a public relations plan. The committee decided that, if the test fizzled, the government would announce that an ammunition depot had exploded. If the test worked, the president would announce the existence of the atomic bomb and its intention to seek control over the new form of military power. Laurence was to draft the statements; Arthur W. Page, an AT&T executive who was assisting with the government's public relations strategy, would review Laurence's work and submit the releases to the committee.[22]

Even before the Interim Committee acted, Groves had summoned Laurence to his office, and he and his deputy, Gen. Thomas Farrell, briefed the reporter on plans for the upcoming Trinity test. A map of New Mexico hung on the wall behind Groves's desk, with Alamogordo circled. Groves instructed Laurence to prepare press releases for four possible outcomes of the Trinity test: a loud noise but little more, some property damage but no deaths, significant property damage and some deaths, and extensive damage and many deaths.[23] That fourth Trinity press release could have been Laurence's own obituary; the government did not declassify any of the four until 1958.

The first two categories were no challenge for a writer of Laurence's talent and dramatic flair. "I just had to explain an explosion heard for many miles and a large flash of light, then a somewhat larger explosion with some property damage," Laurence told an interviewer two decades later.[24] Thus, for the least disruptive case, Laurence quoted the commander of the Alamogordo Air Base saying that "a remotely located ammunition magazine containing a considerable amount of high explosive exploded. There was no loss of life nor injury to anyone and the property damage outside of the explosives magazine itself was negligible." For the next most serious case, in which the Trinity explosion was powerful enough to force some evacuations, the release quoted the commander as admitting a detonation of "high-power explosives" but added the following falsehood: "Weather conditions affecting the content of gas shells exploded by the blast made it desirable to evacuate some civilians from a small nearby inhabited area."[25]

The third false release was a bit more challenging because it involved deaths. This time, the commander would announce that "the premature explosion of material intended for use as an improved war weapon resulted today in the death of several persons on the reservation, including some of the scientists engaged in the test." Laurence even left a spot for the names of the killed to be added to the release.

But the fourth release was even more difficult, because that scenario combined extensive property damage with deaths of important American scientists. "How could I explain why these men were all out in the desert?" he asked himself.[26] Laurence's solution: "The blast heard today in Albuquerque and other communities within a range of ___ miles of the Almogordo [*sic*] Air Base, which caused serious damage in some sections of the communities in the area involved, was due to a premature explosion of quantities of material being tested for use in improved war weapons against Japan."[27]

Years later, nuclear critics interpreted Laurence's false releases as the kickoff of decades government disinformation and misinformation about nuclear technology. The journalist Karl Grossman saw the drafting of the Trinity press releases as "a first major assignment for Laurence: figuring how to mislead the press—and public" about the risk of atomic energy.[28] Corinne Browne and Robert L. Munroe offer a similar critique, arguing that "Laurence's reports were the backbone of the writing, reporting, filming and editing that constituted a yea-saying to nuclear energy throughout three decades."[29]

The archival record contains no suggestion that Laurence had any qualms about providing a false cover story for the government. This is consistent with his willingness to bend, or break, journalistic ethical norms to work for the government in both the surgeon general's office and the Manhattan Project. As Laurence was writing his releases, the Nazis had been defeated, but the war against Japan seemed far from over; likely in Laurence's view, an Allied victory in the Pacific was important enough to justify his actions. Certainly, Laurence's colleagues at the *Times* afterward had no illusions or qualms about the nature of Laurence's work on the Trinity releases. Years later, his newsroom colleague Gelb wrote, with no hint of criticism, that Groves had instructed Laurence "to concoct a false news story" about the Trinity test.[30]

After writing the Trinity releases, Laurence also worked on a statement by Truman to be used after the first atom bombing and a list, requested by Groves, of twenty-nine other ideas for stories that Laurence could write for the Manhattan Project. He submitted both on May 17. The science historian Alex Wellerstein points to the list as evidence that Laurence was complicit in Groves's attempts to steer journalists' attention in certain directions—and away from others: "The goal was that these stories could be distributed to all newspapers, free of charge and without a requirement for attribution, so that the first week (at least) of the news cycle about the atomic bomb would be both extensive and dominated by a Groves-approved narrative. Groves' 'Publicity' strategy was in evidence: selectively release a lot of new, once-secret information, and thereby control what information was available." Wellerstein offers Laurence the benefit of the doubt as to whether the reporter understood the risk of radiation from the bomb, but the historian Peter Kirstein instead argues that Laurence's story ideas showed he well understood the deadly impact of radiation from the bomb.[31]

Laurence simply could not keep his dramatic side in check as he wrote a radio script for Truman, and consequently it ended up being far too long and not at all presidential in tone. For example, on the second page, Laurence would have had Truman say, "This new development, which brings this 'Cosmic Fire' down to earth for the first time, just as Prometheus, father of civilization, brought ordinary terrestrial fire down to earth from Olympus, marks the opening of a new era in our civilization."[32] Laurence would have had the president acknowledge that the defeated Germans had run a substantial nuclear weapons program, explain the differences between various isotopes of uranium and the difficulty in separating them into bomb-grade material, and describe in detail the industrial complex that the United States had erected to manufacture the bomb. The script praised the press for refraining from publicizing the projects, workers at the sites who didn't know the ultimate purpose of their work, and scientists. He had Truman pledge to support a vibrant postwar program in developing atomic power for both military and civilian purposes, including the development of thorium as a power source.

James B. Conant, a prominent chemist and former Harvard president who had known Laurence as an undergraduate at Harvard and was a senior government adviser on wartime technical affairs, didn't like Laurence's statement for Truman. He wrote to Vannevar Bush, "I only had time to glance at it, but it seemed much too detailed, too phoney and highly exaggerated in many places." But Conant reassured Bush that Arthur Page could fix the problem.[33] On May 18 the Interim Committee approved Laurence's Trinity press releases but found that his presidential statement was far too long, and they asked Page to revise it.[34]

Page basically tossed Laurence's work into the trash, writing a new statement for Truman. The four-page statement started, "Sixteen hours ago an American airplane dropped one bomb on Hiroshima, an important Japanese Army base. That bomb had more power than 20,000 tons tons [*sic*] of T.N.T. It had more than two thousand times the blast power of the British 'Grand Slam' which is the largest bomb ever yet used in the history of warfare." Page's statement for Truman went on to say that the bomb is "a harnessing of the basic power of the universe." The statement briefly sketched the efforts to produce the bomb and warned the Japanese that they must now surrender. "If they do not now accept our terms they may expect a ruin of rain form [*sic*] the air, the like of which has never been seen on this earth."[35]

As a science writer, Laurence wanted to devote significant attention to the scientists involved in the Manhattan Project, but the Truman administration pushed back. On June 19, 1945, Lt. Col. William Consodine wrote of Laurence's early drafts: "There is too great a stress on the scientists as a whole. The job was finished because of the complete coordination of the Army, industry and the scientists. There is too little mention of the part played by the Army. I believe that industry has been given proper attention and that the scientists have been given absolutely too much attention."[36]

Laurence's recollection about his role in the presidential statement skirted criticism of his work. In his retelling, his Truman statement was fine, but because Truman was at sea when the bomb was dropped on Hiroshima, officials had to divide Laurence's radio broadcast into two portions, one for Truman and another for Stimson.[37] In 1967 Laurence again misstated matters, this time in a chapter he contributed to a book titled *How I Got That Story*, in which various journalists recounted their professional triumphs. Laurence wrote about the press releases distributed on the day of the Hiroshima bombing, "With minor revisions, everything given out that day by the White House and by the War Department was material prepared by me in advance."[38] Groves also exaggerated Laurence's role, saying that Laurence's proposed statement for Truman underwent only "minor revisions" after Trinity.[39] Not surprisingly, some historians and other scholars have been misled or confused about what Laurence did and didn't do.

As the wordsmithing of Truman's announcement continued, Laurence pursued his tours of Manhattan Project facilities so that he could write other releases that described the atomic bomb effort in greater detail. Laurence estimated that he traveled 50,000 miles by airplane during his time working for the Manhattan Project.[40]

In early June 1945 the Allies agreed to partition Germany into areas controlled by the Soviet Union, United States, Britain, and France, an arrangement that would persist for decades. At the same time, Laurence visited the sprawling Hanford nuclear reservation in Washington State, which was home to nuclear reactors that produced plutonium for weapons use. He arrived there on June 1 and ate up the details of the operation. R. Monte Evans, an official from DuPont, which ran Hanford for the government, immediately asked to meet with the reporter, and he provided Laurence an extensive tour of Hanford's 300 Area, where raw uranium was fashioned into fuel rods for insertion into the plutonium-breeding reactors. Laurence also viewed a demonstration of a crane and remote control apparatus in Hanford's Building 221U, which was designed to chemically extract plutonium from the fuel rods after they had been removed from the nuclear reactors and were highly radioactive. When Laurence visited, Building 221U had been completed but was being kept in a standby condition; the building never was used for plutonium production. His visit to Hanford left a huge impression on the reporter, who later recalled: "I shall never forget the feeling of awe I experienced when I first saw in the semidesert of Washington State the gigantic plants known as the Hanford Engineering Works, where the plutonium for atomic bombs was being created by modern alchemy."[41] Laurence departed Hanford on June 9 after telling officials there that "he had sufficient information for his work to write a book."[42] This comment indicates that Laurence was already plotting ways to leverage his Manhattan Project experiences in the postwar years. In his releases, his writing for the *Times*, and his books, he would refer to Hanford as "Atomland-on-Mars," a mash-up of his lifelong fascinations with the atom and the planet Mars. He described Hanford as "a scientific Never-Never land, where the accepted 'impossibles' of yesterday had become actualities of staggering dimensions, in both space and time."[43]

Laurence's arrival at the main weapons design laboratory in Los Alamos, New Mexico, on a Saturday in May 1945 turned heads, including that of Dorothy McKibbin, who managed access to the facility with a pleasant but ironfisted demeanor. "At first, no one could believe it. As far as Dorothy knew, he was the first and only reporter ever invited up to the classified weapons installation," one historian noted.[44] Laurence caused a further ruckus when he appeared at a party in Los Alamos. "Most of the scientists at the party recognized him and several rushed to Colonel Gerald Tyler, head officer in charge of security, with the information that there was a reporter from the *New York Times* present. Colonel Tyler calmly told them that he knew all about it," one scholar related. The scientists then were briefed about Laurence and his role in the Manhattan Project.[45]

Senior scientists were particularly helpful to Laurence. Conant told Laurence, "They won't believe you, when the time comes that this can be told. It is more fantastic than Jules Verne."[46] Oppenheimer also talked with Laurence. In Oppenheimer's office at Los Alamos, the physicist opened a safe and removed a vial of superheavy water, or water made of oxygen and tritium, a radioactive form of hydrogen used in atomic weapons. "We both looked at it in silent, rapt admiration," Laurence later wrote. "Though we did not speak, each of us knew what the other was thinking. . . . Here was something with the power to return the earth to its lifeless state of two billion years ago."[47]

Given the latitude that Groves had granted Laurence, some researchers were willing to brief him on topics not directly related to the war effort. The chemist Glenn T. Seaborg prepared Laurence a seven-page report, classified secret, on the discovery of uranium-233, an isotope of uranium that was not being used in the current weapons, and the early research into its properties. (The copy of the report in Seaborg's papers at the Library of Congress bears the handwritten notation, "Written for Laurence.") Seaborg's document would have been a treat for Laurence to read, providing a description of how the isotope could be produced by bombarding thorium with neutrons. Among other points, Seaborg's report emphasized the possibility of constructing a breeder reactor that would incorporate thorium along with uranium-233. The thorium would absorb some of the neutrons released by the fission of uranium-233 atoms, and in so doing the thorium atoms would be transmuted into uranium-233 atoms. The process would generate more uranium-233 atoms than it would consume. "This makes thorium potentially as important as uranium in the atomic power development," Seaborg wrote.[48] Laurence would pick up on this theme, emphasizing the importance of thorium in his Manhattan Project press releases, to the irritation of some of those reviewing his work for distribution; for example, in editing one of Laurence's Manhattan Project releases, Consodine questioned "whether it is necessary to mention thorium at all in the article."[49] Laurence similarly touted thorium in his postwar reporting on atomic power. So-called breeder reactors would be proposed, but meet with public opposition, in later decades, but Seaborg probably initiated Laurence's excitement for the concept with this report.[50]

Although Laurence found all the information about the Manhattan Project tremendously exciting, he also had trouble absorbing and processing it all. Laurence well knew that his writing would be censored before publication, but he sought to do as thorough and complete a job as possible. "I had a free hand completely to ask any question that I wanted, to go to any place that I wanted, to see everything for myself, and then to write as good an accurate and detailed a report, and, again, as entertaining and lively a report as I knew how."[51] He gathered information during the day and wrote at night at his office at Oak Ridge, sometimes very late. He tore his drafts into small pieces and tossed them into a red metal wastebasket with the word BURN on it. Two guards burned the contents every day. "I learned after a while that these particular guards had been carefully chosen after it had been determined that both were illiterate."[52]

One day, Laurence tried to bring a copy of his own 1940 *Saturday Evening Post* article about the atom into the Oak Ridge facility, but he was told that he couldn't bring it in because it was classified. " 'Classified?' asked Laurence. 'At least twenty million people have seen it, I hope,' " the magazine crowed after the bombs were dropped.[53] Laurence also later learned that a copy of the article, along with a German-language translation of it, were found in a captured German laboratory.[54]

"It was the toughest assignment I've ever covered, but it was the greatest experience any newspaper man ever could wish for," Laurence told the industry publication *Editor and Publisher* after the war's end.[55]

Groves's staff and ultimately Groves himself thoroughly edited Laurence's other proposed press releases. Laurence submitted his press releases to Consodine, who farmed them out to Lt. Col. Clyde H. Matthews and Capt. Kilburn R. Brown. If Laurence assumed they would defer to a Pulitzer Prize winner, he was wrong. The officers rewrote Laurence's pieces, sometimes multiple times. The ultimate editing was left to Groves himself. A set of drafts of Laurence's press releases, housed in the National

Archives in College Park, Maryland, bear editing marks that apparently were made by Groves.[56]

Media scholar Beverly Keever found that the War Department censors eliminated references in Laurence's releases to concepts such as doomsday, the Promised Land, and moon rockets; changed references to secrecy into references to security; edited text to play down the risk of radioactivity; deleted references to resistance to government seizure of property to develop facilities such as the Oak Ridge complex; and downplayed the role of labor, particularly the thousands of women who worked in the Manhattan Project. Keever concludes, "Laurence's drafts . . . illuminate some facets of A-bomb-making that rarely appeared in news articles of the time. And, in turn, the Laurence drafts also etch out a U.S. government policy that shaped and controlled the first information about this revolutionary weapon before it was announced to the world."[57]

In early June, Groves's staff was considering how to orchestrate the announcement of the atomic explosion against Japan. Because some newspapers were published in the morning and others in the afternoon, the choice of an announcement time inevitably would disadvantage some newspapers in covering the historic story. The decision was to time Truman's announcement to enable afternoon papers to trumpet the news. Later that day, more details—including Laurence's releases—would be distributed, giving new material for the next day's morning newspapers.[58] In all, the MED expected to distribute twenty press releases, eight by Laurence, five by public relations officer Maj. John F. Moynahan, and seven by Robinson.

By July 12, days after the liberation of the Philippines, Laurence had wrapped up his research on the development and details of the bomb and was ready to move on to the Trinity test, which would happen four days later.[59] "I had seen things no human eye had seen before—that no human mind before our time could have conceived possible. I had watched in constant fascination as men worked with heaps of uranium and plutonium great enough to blow major cities out of existence. I had prepared scores of reports on what I had observed, and I couldn't help but dream wistfully of what sensations they would have been in The Times," he later told Meyer Berger.[60]

As Laurence prepared to leave Oak Ridge for the Trinity test on July 16, Laurence wrote James a letter. Without providing specifics, Laurence warned his editor at the *Times* that the story he was working on would be a blockbuster. It is unclear to this day whether Laurence had permission to disclose even that much information or how Groves would have regarded it.[61] "The story is much bigger than I could imagine, fantastic, bizarre, fascinating and terrifying," Laurence wrote James. "When it breaks it will be an eighth day wonder, a sort of Second Coming of Christ yarn. It will be one of the big stories of our generation and it will run for some time. It will need about twenty columns on the day it breaks. This may sound overenthusiastic, but I am willing to wager you right now that when the time comes you will agree that my estimate is on the conservative side." Laurence predicted to James that he would be back in the newsroom between September 15 and October 1. "Even my identity is kept a deep secret and I find myself slinking around corners for fear someone may recognize me." Laurence promised that his involvement would pay off for the *Times*: "After the story breaks I will be the only one with first hand knowledge of it, which should give The Times a considerable edge. Much of it, however, will be kept on ice for some time."[62]

When James received the letter, he sealed it in an envelope and passed it to Sulzberger. "I thought you would like to read this letter from Laurence. Of course, it is very confidential." Sulzberger responded: "Thank you. This looks like IT."[63]

5

Trinity, Hiroshima, and Nagasaki

Bracing for history to be made, Laurence returned to Los Alamos after mailing the letter to his editor. When arriving at Los Alamos, Laurence was taken aback by the pessimism that he encountered among the physicists. The plutonium bomb was projected to have an explosive energy equal to that of 20,000 tons of TNT, but some of the Los Alamos scientists doubted that the bomb would explode at all. Laurence later recalled,

> I remember only too well the atmosphere of doubt that pervaded the scientists up to the last minute of the test. On the one hand, they were haunted by the fear that their "gadget," the product of the greatest concentration of brainpower in history, might turn out to be a "fizzle," which meant either a complete dud or the equivalent of no more than a few ordinary blockbusters in terms of T.N.T. On the other hand there were many, including some of the elite among them, who feared that man, like Sorcerer's Apprentice, may have started something beyond his control.[1]

The Los Alamos scientists had started a betting pool about how powerful the Trinity test would be, and many of the scientists put their money on low numbers.[2] Oppenheimer, the project's visionary lead scientist, selected only 300 tons of TNT. To buck the trend, Laurence paid one dollar to bet on 13,700 tons. Trinity actually equaled 21,000 tons of TNT, and the pool was won by the physicist I. I. Rabi, who had bet on 18,000 tons.[3]

The Trinity test was carried out in Alamogordo, New Mexico, about 200 miles south of Los Alamos. Laurence and a few others left Los Alamos on the afternoon of July 15, planning to stop in Albuquerque on the way. Accounts differ on the details. Laurence told the author Lansing Lamont that he drove to the Trinity test with Sir James Chadwick and Edwin McMillan, with Col. Gerald R. Tyler, the post commander at Los Alamos, catching a ride as far as Kirtland Air Field (today called Kirtland Air Force Base) there. In Laurence's account, the car mates debated the extent to which the Americans should share classified atomic information with the Russians and arrived in Albuquerque about noon on Saturday. But according to another account, Laurence rode with Chadwick and Ernest O. Lawrence. Moreover, Chadwick later independently stated that his conversation with Laurence happened after the Trinity test, in a car ride back to Los Alamos.[4]

In Albuquerque they had to kill several hours before continuing on to Alamogordo. "We arrived there, and we sort of didn't know what to do. We were standing around the streets, and so forth, on corners," Laurence recalled.[5] Always worried about security, Groves was horrified to see a large clump of his scientists gathered in the Albuquerque Hilton, and he ordered them to disperse and make themselves less obvious. Lawrence, Laurence, and some others took refuge in a hamburger stand.[6] Laurence said he and the others "spent most of the afternoon sort of in hiding . . . and standing in little crowds or in side streets, or sitting in the cars or in the buses." Finally the caravan left for the Trinity Site at about 11:00 p.m.[7]

Because the Trinity explosion initially was scheduled for 4:00 a.m. on July 16, the physicists and other observers drove through the night to get there in time, arriving at Alamogordo about 3:00 a.m. The bomb had been hoisted to the top of a 100-foot steel tower in the middle of the desert, and the Trinity team fanned out. Some team members were scattered among three observation sites about two miles from the tower. Groves

Figure 5.1. The plutonium device tested at Trinity was called the Gadget. Sitting next to it is Norris Bradbury, head of the bomb assembly group. (US Department of Energy)

was at base camp about ten miles from ground zero. Laurence and VIPs were dispatched to Compania Hill, about twenty miles north of ground zero; historians today are uncertain of the exact location.

Like any reporter, Laurence coveted an up-close-and-personal view of Trinity, so he was peeved to be stuck so far from it on Compania Hill. And despite his bet on 13,700 tons of TNT as the bomb's strength, he apparently didn't realize how far from ground zero the devastation would extend if the bomb had the explosive energy that he had predicted. Laurence complained to Groves, who replied tartly, "You'll see plenty."[8]

The weather was not promising. "It had been raining, and was completely overcast," Joseph W. Kennedy, an assistant to Seaborg, wrote in his diary. Like Laurence, he was assigned to Compania Hill. "Finally we learned by MP radio that the shot would be delayed (due to weather) for perhaps one or two hours." The desert night was cold. Laurence and

Figure 5.2. The Gadget was hoisted atop a 100-foot steel tower anchored into the New Mexico desert, in order to reduce fallout and aid scientific measurements of the explosion. (US Department of Energy)

the others drank coffee and ate sandwiches. "We sat around and just talked and talked," Laurence recalled. "Various ones tried to sleep on the ground," Kennedy further noted in his diary. "CAT [chemist Charles Allen Thomas], Laurence and I tried in the sedan, but L. [Laurence] snored loudly."[9]

Others at Compania Hill tried to while away the time. A Los Alamos history recalls, "They shivered in the cold and listened to instructions read by flashlight by David Dow, in charge of that observation post. They ate a picnic breakfast. The Hungarian-born physicist Edward Teller warned about sunburn and somebody passed around some sunburn lotion in the pitch darkness."[10] As the moment of detonation approached, Lansing Lamont wrote in his history of Trinity, "Bill Laurence wetted the tip of a pencil and brought out his note pad."[11]

The light from the blast illuminated the night sky for miles and caused alarm among some New Mexico residents who worried that a major weapons accident had happened. Just as a lightning flash precedes a thunderclap by a few moments, so the light from Trinity was followed by a booming sound. "It was the blast from thousands of blockbusters going off simultaneously in one spot. The thunder reverberated all through the desert, bounced back and forth from the Sierra Oscuros, echo upon echo," Laurence later wrote for the *Times*.[12] About a year after Trinity, Laurence recalled the roar of the explosion in a radio interview while circling the Trinity Site in an airplane. "When the sound finally came, it felt as though the Earth itself had split apart, and none of us would have been surprised if we had all started falling through toward the South Pole."[13]

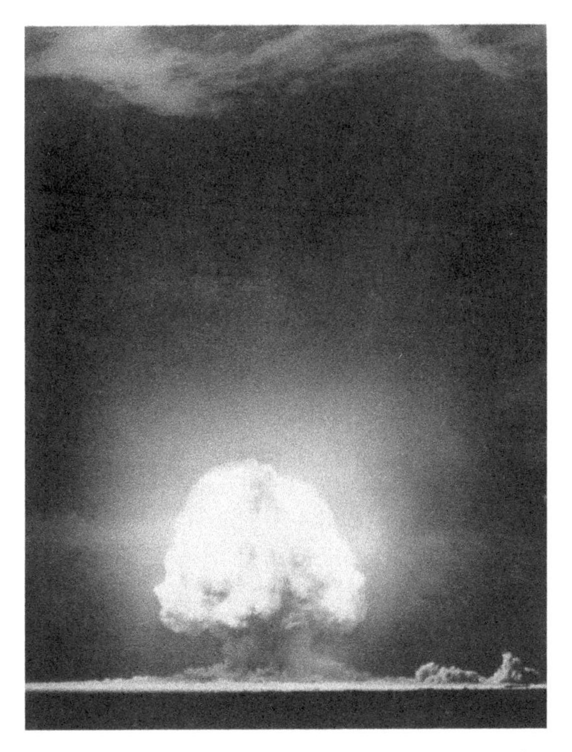

Figure 5.3. The Trinity explosion in the predawn darkness of July 16, 1945, demonstrated that the design of the plutonium-fueled weapon was correct. Laboratory data had already shown that the other design produced by the Manhattan Project, using uranium, would work without need of a field test. (US Department of Energy)

But on the morning of the Trinity blast, the noise at first perplexed Laurence. The physicist Hans Bethe recalled, "Laurence was terribly afraid and cried out, 'WHAT WAS THAT?' So I explained to him that sound takes some time to propagate as compared to light."[14] The physicist Victor Weisskopf recalled Laurence more charitably in his own memoir, explaining that Laurence "asked, 'What was that?' because, having taken so long to reach us, the sound seemed unconnected to the explosion."[15] Yet another Trinity chronicler, Stephane Groueff, wrote that Laurence realized what the noise was on his own, mumbling after his question: "Oh, that was the bomb we've just been watching."[16]

Laurence's reaction to the bomb's thunderous noise was well known among the Trinity physicists. "We all joked about that," the physicist Robert Marshak recalled thirty years later.[17] But another physicist, Jeremy Bernstein, suggests any mockery was unfair. "He was actually asking a rather good question," because of the long delay between the flash and the sound, Bernstein wrote.[18]

Over the years, Laurence wrote at least four distinct accounts of the Trinity explosion. The first was a press release that was part of the package of materials distributed by the government after the Hiroshima bombing. The second was his account that he wrote for the *Times*, published in the paper more than a month after Hiroshima and included almost verbatim as chapter 1 of *Dawn over Zero*, his first book, about a year after Hiroshima. The third was also included in *Dawn over Zero*, as a later chapter. Laurence's fourth description of Trinity was included in his third book, *Men and Atoms*, published in 1959. The progression of treatments shows escalating hyperbole and drama as well as numerous errors of fact and interpretation.

For example, his description of the Trinity explosion in the press release was remarkably understated—for Laurence. "At the appointed time, there was a blinding flash lighting up the whole area brighter than the brightest daylight. A mountain range three miles from the observation point stood out in bold relief." His account for the *Times* and *Dawn over Zero* demonstrated more of his trademark drama: "And just at that instant there arose from the bowels of the earth a light not of this world, the light of many suns in one. It was a sunrise such as the world had never seen, a great green super-sun climbing in a fraction of a second to a height of more than 8,000 feet, rising ever higher until it touched the clouds, lighting up earth and sky all around with a dazzling luminosity."[19] He repeated the same description of the blast in *Men and Atoms*, thirteen years later.

In his press release, Laurence depicted the scientists reacting with excitement and relief, writing that Drs. Conant and Bush had both shaken hands with Groves after the blast. "All the pent-up emotions were released in those few minutes," Laurence wrote. In his *Times* story, Laurence was much more descriptive. "A loud cry filled the air. The little groups that hitherto had stood rooted to the earth like desert plants broke into a dance, the rhythm of primitive man dancing at one of his fire festivals at the coming of spring. They clapped their hands as they leaped from the ground—earth-bound man symbolizing a new birth in freedom—the birth of a new force that for the first time gives man means to free himself from the gravitational pull of the earth that holds him down."[20]

Laurence saw the Trinity test as marking the start of a new epoch in human history, an event with such wrenching emotional power that it could inspire its witnesses to spontaneously render deep philosophical reactions. In both his *Times* story and *Dawn over Zero*, Laurence described Harvard physicist George B. Kistiakowsky, a future presidential science adviser, to have declaimed after Trinity, "I am sure that at the end of the world—in the last milli-second of the earth's existence—the last man will see what we saw!"[21] But Kistiakowsky later contradicted Laurence's reporting, telling Lamont that he had made the statement two days after the Trinity test, in a Los Alamos cafeteria. "When you're very nervous and tense you don't make historic statements like that," the physicist said. "You say something silly."[22] In *Men and Atoms*, Laurence changed his timeline and portrayed Kistiakowsky as speaking in a Los Alamos dining room at breakfast on Monday after the Trinity explosion.

Laurence's 1959 book *Men and Atoms* introduced another new element to Laurence's Trinity narrative. In it, for the first time, Laurence reported that Oppenheimer had made a historic statement shortly after the Trinity explosion. Laurence wrote in the book, "'At that moment,' I heard him say, 'there flashed into my mind a passage from the *Bhagavad Gita*, a sacred book of Hinduism: "I am become Death, the Shatterer of Worlds!"'"[23] Laurence was not the first to report the *Gita* quote, and historians do not dispute that Oppenheimer cited the *Gita*. In fact, much analysis has centered on what Oppenheimer meant by what he said, not whether he said it. Of interest here is the fact that the *Gita* reference only appears in Laurence's writing fourteen years after Oppenheimer's statement, which suggests that Laurence may have been cribbing from others' reporting without citation. The historian James A. Hijiya argues that *Time* appears to have been the first to report the quote, in November 1948—after Laurence's initial writings on Trinity. *Brighter Than a Thousand Suns*, the 1958 book on Trinity by Robert Jungk, contains a longer version of the quote, which Hijiya suggested could be a "later extrapolation" by Oppenheimer.[24] Bird and Sherwin suggest that Laurence drew the quote from *Brighter Than a Thousand Suns*. If so, Laurence did not footnote or otherwise credit Jungk.[25]

Laurence's penchant for drama got him into hot water with Kenneth Bainbridge, a Harvard physicist who was working on the bomb. Bainbridge complained that Laurence's official release on the Trinity test contained "many errors and misstatements."[26] In *Dawn over Zero*, Laurence emphasized officials' worries that poor weather would delay the Trinity test. In the book, Laurence portrayed Bainbridge, Kistiakowsky, and Howard C. Bush (the head of Trinity's military police) pacing atop the tower that held the test bomb: "These three stood watch on top of the tower from one o'clock until a half hour before zero, their silhouettes outlined at intervals by a flash of lightning."[27] Bainbridge later wrote that the scene was "purely imaginary. Laurence was 20 miles away."[28]

That morning, a few reporters inquired about the overnight explosion. At about 11:00 a.m., the Alamogordo Air Base issued Laurence's second-level Trinity press release—which cited the need to evacuate some civilians after an accident involving high explosives and pyrotechnics. The newspaper in El Paso, Texas, reported that "many persons saw a flash light up the sky, like daylight, and felt earth tremors."[29] At 4:46 p.m. on July 16, the United Press distributed a story quoting the air base commander as saying that an ammunition magazine had exploded, including some gas shells, and weather conditions might necessitate some evacuations. The Office of Censorship approved the story but advised that evacuations were not going to be necessary after all. The United Press reporter seems to have realized that he was being fed a cover story, because he then asked the censorship official whether "a lot of people were running around out there with tears in their eyes." The official allowed that it was a "hot story" but said he "could not talk about it."[30]

The Trinity blast was also reported by the Associated Press, and its article appeared in papers as far as Albuquerque and Tucson. East Coast papers did not publish the article, which was distributed on a regional AP circuit rather than nationwide.[31] "The judgment of the Press Division proved good," the censorship office later congratulated itself in its official history. "Next morning, in the welter of big news, stories on the New Mexico explosion were virtually lost in the nation's newspapers."[32] (However, the agency did bar the information from being sent overseas.)

After the explosion, Laurence arrived back at the Albuquerque Hilton at around 10:00 a.m. and cleaned up.[33] In a 1964 oral history Laurence stated that he went straight to Los Alamos and immediately typed his report about Trinity, but in *Men and Atoms* he described his writing much more metaphysically and timed it as happening on Monday: "Without full awareness, I took out my notebook and pencil and began writing feverishly, like one waking from a dream, in a frantic effort to record it before the return of full consciousness blotted it from my memory."[34]

Laurence was eager to get to the Pacific to witness the first use of the bomb in war. But first he briefly returned to Oak Ridge, his home base while working for the Manhattan Project. While there, he visited the home of Oak Ridge scientist Jerry R. Coe. Richard Gehman—a journalist and novelist who was serving as editor of the *Oak Ridge Journal*, the secret town's government-sanctioned newspaper—also dropped by, much to Laurence's surprise. Gehman later described the post-Trinity Laurence as wearing "the look of a man, privileged to peek through a crack in the wall that hides the mystery of the universe, confounded by what he had seen, awed in contemplation of his own insignificance, wondering where it will all lead. It was a look I will never forget."[35]

On July 15, the day before Trinity, President Harry S. Truman and his senior aides had arrived in Potsdam, near Berlin, to develop a plan with the other Allies for the postwar era. After the explosion, Secretary of War Henry Stimson received a telegram from an aide with a preliminary report: "Operated on this morning. Diagnosis not yet complete but results seem satisfactory and already exceed expectations. Local press release necessary as interest extends great distance. Dr. Groves pleased. He returns tomorrow. I will keep you posted." Two days later Stimson received a follow-up report: "Doctor has just returned most enthusiastic and confident

Figure 5.4. Laurence on Tinian, an island in the Northern Mariana Islands that the United States used as its base for dropping the atomic bombs on Hiroshima and Nagasaki. (National Museum of Nuclear Science and History)

that the little boy is as husky as his big brother. The light in his eyes discernible from here to Highhold and I could have heard his screams from here to my farm." The code was meant to express the reassuring news that the plutonium device used at Trinity appeared to be as powerful as the uranium device, which had not needed testing.[36]

With such results, the use of the atom bomb in the war was finally at hand, and Laurence had no intention of missing that story. However, compartmentalization reigned supreme. Laurence recalled, "On my return from the bomb test in New Mexico, I was not permitted to tell even Colonel Nichols what I had seen."[37]

Groves insisted that Laurence visit his wife Florence before he traveled to the Pacific. Florence had emigrated from Lithuania to New York in 1922 and nine years later married William. Laurence later told

Meyer Berger that Groves instructed him, "Mrs. Laurence will want to know where you're going. Tell her you're going to London, but that you don't know what for. Clean out your desk first. Burn anything you have around."[38] Laurence was accompanied on his visit to his home by an FBI agent. Laurence told Florence that he was a coworker, but when the guard took off his coat, two guns showed. He couldn't tell his wife what he was up to or where he was going. "It was a rather unsatisfactory farewell," Laurence recalled.[39]

His overseas trip started at Hamilton Field near San Francisco, which was overrun with generals holding higher priority for transportation than he did. He cooled his heels there for three days before speaking up to the management, which finally got him onto an army transport bound for Honolulu. An exhausted Laurence fell asleep after takeoff and woke up as the plane was descending to land—but back at San Francisco rather than Honolulu, because the plane had developed mechanical trouble while he was sleeping and was forced to return to its departure point. That meant more delay waiting for a seat on another plane; meanwhile, Manhattan Project officials in the Pacific were wondering where he was.

After he landed at Guam on August 5, Laurence opened sealed instructions that told him how to contact staff who would get him to Tinian, a small island about 50 miles north of Guam that the Americans had captured from the Japanese the previous year. Tinian was the operations base for its Twentieth Air Force and its strategic bombing operations against Japan. The 509th Composite Group, a unit of the Twentieth Air Force that had been training to deliver the atomic bombs, was on Guam getting ready for the attack.[40]

When Laurence finally arrived at Tinian, General Farrell broke the news that he was too late to get aboard the Hiroshima flight. "I was in despair. I had missed the Big Chance," Laurence wrote in *Men and Atoms*.[41] Farrell promised Laurence that he would fly on the second mission, but Laurence worried that there might not be a second mission if the first managed to end the war.[42] The Enola Gay was under the direction of Gen. Carl Spaatz, the commander of US Strategic Air Forces in the Pacific, and Gen. Curtis LeMay, so any appeals Laurence directed to Groves were pointless.

It remains unclear why Laurence couldn't go on the Hiroshima attack. Weight issues are often cited. Admittedly, the bomb was unusually heavy, and Col. Paul W. Tibbets Jr. and his crew had been practicing for months on how to safely deliver it to a target. In his account of the bombing, the journalist Joseph L. Marx wrote that Laurence couldn't go on any of the Hiroshima planes because "they were already filled to capacity with men, fuel, and scientific equipment."[43] In a 1964 oral history, Laurence agreed that weight restrictions on the bomber probably were to blame, pointing out that the plane was carrying twelve people compared to its usual complement of nine. If he had arrived in Tinian on time, perhaps he would have gotten one of the seats held by scientists.[44]

However, Groves may have not wanted Laurence on the flight because he would have upstaged other reporters. Around August 3, Farrell proposed that Laurence "write up [the] entire project operation" but added, "It is not proposed that he write spot news for immediate transmission for release in Washington since such action would be in competition with accredited theatre correspondents and violate established procedure."[45]

Groves may never have approved Laurence's inclusion in the Hiroshima attack. After the Hiroshima success, Farrell cabled his superiors on August 7 asking that Laurence be permitted to fly "on combat missions." There would have been no need to seek such orders if Laurence previously had been cleared to fly. The copy of Farrell's cable in the National Archives bears the handwritten notation, "LRG handled by cable to Spaatz," indicating that Groves then reached out to Spaatz.[46] This aligns with a recollection by Robert A. Lewis, the Enola Gay's copilot, who kept a logbook of the Hiroshima attack at Laurence's request. Years later, Lewis said that he had originally written in the diary that a request for Laurence to fly on the Hiroshima mission "was not put in," but Laurence had revised the passage to state "but he arrived in Tinian too late."[47]

Whatever the reason for being excluded from the Hiroshima mission, Laurence tried to find a silver lining in the unique access he had, even while stuck on the ground. He was on Tinian along with the only two assembled atomic weapons in existence and a cadre of scientists and engineers. He had never seen the fully assembled uranium weapon, so he sought out Little Boy, the uranium bomb that was to be used over Hiroshima. "I spent the afternoon of that Sunday inspecting the plane and particularly the bomb that was suspended in its bay." He paid enough attention to details that he was able to recall in a 1964 oral history interview that the bomb contained 66 kilograms of uranium 235 (close to the precise amount of 64 kilograms) and to describe the arrangement of components that fired one piece of U-235 into another piece of the same material, forming a critical mass that exploded.[48]

One scientist at Tinian was the physicist Lawrence M. Langer, who had met Laurence during one of his visits to Los Alamos. When Laurence started asking questions on Tinian, Langer was reluctant to answer, but a security officer reassured Langer that he should tell Laurence anything he wanted to know. Laurence's first question had nothing to do with nuclear technology: "How big are the motors on the aircraft?"[49]

Tibbets didn't want to hassle with Laurence, so he foisted the reporter onto Lt. Jacob Beser, the Enola Gay's radar specialist, who had been planning to sleep before takeoff. "I interpreted this to mean, 'please keep him occupied and out of my hair,'" Beser later recalled. He said that Laurence "had a mind that could comprehend the most complicated scientific theory and reduce it to simple layman's language." But nevertheless Beser limited their conversation to unclassified matters.[50] The two talked until shortly before the midnight flight briefing.[51]

Laurence watched as the bomber lumbered into the air at about 2:45 a.m. "Our hearts were right in our mouths. We were just praying he would get off the ground because if he went a little further he would have cracked up," Laurence recalled.[52] Because it would be hours before the bomber would return to Tinian, many of the ground personnel then headed to bed. Laurence was assigned to share a tent with Langer. "I had never heard anyone snore that loud," the physicist recounted thirty-seven years later. "He just about shook that tent all night long."[53]

While the Enola Gay was en route to Hiroshima, some concern arose about accountability for the uranium that had been produced for the bomb at great expense. The metal in the bomb had an estimated value of $500 million, reflecting the costly industrial processes that had manufactured it.[54] Farrell, Groves's deputy, signed a receipt for a "projectile unit containing [deleted] kilograms of enriched tuballoy at an average concentration of [deleted]." (Tuballoy was the code word for uranium.) Farrell added a handwritten notation, "The above materials were carried by Parsons, Tibbets & Co. to Hirohito's part of 'Doomsday' leaving Tinian at 051645Z." Captain W. S. Parsons added his own annotation: "I certify that the above material was expended to the city of Hiroshima, Japan at 0915 6 Aug." Laurence and nine other civilians and military men also signed the receipt as "witnesses."[55] Decades later, this may seem to have been a frivolous way of frittering away time while waiting for the results of the bombing run. However, precedent existed; before the Trinity test less than a month before, Farrell had signed a receipt for the plutonium used in that explosion.[56]

The Enola Gay dropped its atomic bomb over Hiroshima at 8:15 a.m. local time (9:15 a.m. Tinian time), and Tinian received a coded radio message reporting the bombing a few minutes later. Laurence was present when the Enola Gay landed at 3:00 p.m., and Spaatz immediately pinned a medal on Tibbets. One of the first things Laurence did was retrieve the logbook from Lewis.[57]

With the bombing accomplished, much of the secrecy about the bomb project immediately evaporated. "It was a curious sensation to stand right

Figure 5.5. Laurence posed with public relations officer Maj. John F. Moynahan on Tinian before takeoff of the Hiroshima mission. (US Air Force)

Figure 5.6. The crew of the Enola Gay dropped the Little Boy uranium bomb on Hiroshima at 8:15 a.m. local time on August 6, 1945. (National Archives and Records Administration, Record Group 342, archives identifier 342-AF-58189)

there at the very heart of things, knowing that all the world was just then being electrified, as it were, with the very energy of the substance a few feet away from the radio," Laurence wrote in *Dawn over Zero*. "It was shocking at first to hear terms such as 'atomic energy,' 'uranium 235,' 'atomic bomb' come out openly on the radio. These words had been strictly taboo. They were never uttered even in a whisper. One always talked about such things in code. There were always animated conversations about 'barber shops' and 'pigs,' or we called numbers, like a quarterback calling signals, or the letters of the alphabet."[58]

In New York, Turner Catledge, who had ascended to the post of assistant managing editor at the *Times* in January, already knew that big war news would be coming soon. After Laurence had departed on his secret mission in May, James informed Catledge that Laurence was working on a classified project for the army. In his memoir, Catledge later described how he was initiated into the secret of the atomic bomb:

> One day in August of 1945 a man I knew from Washington, a Mississippian who was an Army captain but who worked in civilian clothes, cames [*sic*] to see me in New York. He sat down beside by [*sic*] desk and asked if I knew what Bill Laurence had been doing. I said I did not, and he replied that I would soon find out. He instructed, or rather ordered, me to report to a certain building in Washington the next day. He said I should tell Sulzberger I was going to Washington on "the Laurence matter" but should say nothing to anyone else.[59]

When Catledge arrived, he found himself across a conference table from Groves, who briefed Catledge on the bomb, the Manhattan Project, the Trinity test, and the plan to drop a bomb on Japan. "They were taking the *Times* into their confidence to ensure that we would print a full and accurate account of the bomb's use and development when the time came," Catledge recalled. "I was told almost everything except when and where the first bomb would be dropped."[60] According to the corporate history of the *Times*, Groves told Catledge, "We've called you down here because we've taken your science man away, and this is strictly a science story. We feel a little under obligation to The Times, so I'm taking this way to give you a little edge. I just want to give your paper a fair break; get you alerted for an important story."[61]

Catledge warned Sulzberger, who summoned Kaempffert back to town so he could write a science editorial if appropriate.[62] Catledge assigned the news story to reporter Sidney Shalett, a reporter familiar to Groves, and the Washington bureau manager placed Shalett on twenty-four-hour availability for the story.

Officials at the government's censorship offices had been warned at 11:00 p.m. the evening before to have a post-bombing announcement ready. They knew about the atom bomb project and in early 1945 had taken to referring to the bomb as the "washing machine" or "Lansdale's lollypop," a reference to Col. John Lansdale, chief of security operations in the office.[63]

At the White House on August 6, Eben Ayers, the acting press secretary, warned reporters at 10:30 a.m. that there might be an announcement later that day. "It'll be a pretty good story," Ayers allowed. A half hour later, he gathered the press corps, read part of Truman's announcement about the bombing of Hiroshima sixteen hours previously, and handed out copies of the full statement. United Press sent the story out a few minutes later. Initially, the White House issued only statements by Truman and Stimson and a brief War Department statement, which perhaps explains why the first coverage by journalists was tentative at best. For example, one radio commentator speaking two hours after the announcement started his dispatch, "This has been an eventful morning. President Truman announced that an atomic bomb had been dropped over Japan, and earlier, Senator Hiram Johnson of California died. Senator Johnson was a man who . . ."[64] Hours after the White House's announcement, the War Department distributed a package of fourteen releases with background information, most of which Laurence had written.[65] The package included a note: "Much of the scientific background of these releases was prepared by William L. Laurence, Science Writer of the New York Times. A Pulitzer Prize Winner and one of the first to write of the possibilities of atomic energy, Mr. Laurence was granted leave by the Times to come with the Project."[66]

At about 11:00 a.m. on August 6, just like other news organizations, the *Times* learned that the bomb had been dropped over Hiroshima. The paper's staff scrambled. Kaempffert wrote a long editorial remarking on the role of science in the development of the weapon, stating, "The bomb which has wrought such havoc on the Japanese Army base of Hiroshima and which, weight for weight, is 2,000 times more destructive than any explosive previously compounded is undoubtedly the most stupendous military and scientific achievement of our time. It may even be the most stupendous ever made in the history of science and technology."[67] On page 5, the paper also ran a story noting Laurence's role. "Behind the pounds of official reports and bales of War Department 'handouts' designed to enlighten laymen on the working of the atomic bomb that was used for the first time over Japan yesterday lay several months of labor by an unassuming New York Times reporter, William L. Laurence," the article said. But the brief, unbylined article offered no detail; other than

"All the News That's Fit to Print"

The New York Times.

LATE CITY EDITION

VOL. XCIV..No. 31,972. NEW YORK, TUESDAY, AUGUST 7, 1945. THREE CENTS

FIRST ATOMIC BOMB DROPPED ON JAPAN; MISSILE IS EQUAL TO 20,000 TONS OF TNT; TRUMAN WARNS FOE OF A 'RAIN OF RUIN'

HIRAM W. JOHNSON, REPUBLICAN DEAN IN THE SENATE, DIES

[illegible]

Jet Plane Explosion Kills Major Bong, Top U. S. Ace

Flier Who Downed 40 Japanese Craft, Sent Home to Be 'Safe,' Was Flying New 'Shooting Star' as a Test Pilot

[illegible]

KYUSHU CITY RAZED

Kenney's Planes Blast Tarumizu in Record Blow From Okinawa

REPORT BY BRITAIN

'By God's Mercy' We Beat Nazis to Bomb, Churchill Says

Steel Tower 'Vaporized' In Trial of Mighty Bomb

Scientists Awe-Struck as Blinding Flash Lighted New Mexico Desert and Great Cloud Bore 40,000 Feet Into Sky

[illegible]

NEW AGE USHERED

Day of Atomic Energy Hailed by President, Revealing Weapon

HIROSHIMA IS TARGET

'Impenetrable' Cloud of Dust Hides City After Single Bomb Strikes

WASHINGTON, Aug. 6—The White House and War Department announced today that an atomic bomb, possessing more power than 20,000 tons of TNT, a destructive force equal to the load of 2,000 B-29's and more than 2,000 times the blast power of what previously was the world's most devastating bomb, had been dropped on Japan.

The announcement, first given to the world in utmost solemnity by President Truman, made it plain that one of the scientific landmarks of the century had been passed, and that the "age of atomic energy," which can be a tremendous force for the advancement of civilization as well as for destruction, was at hand.

At 10:45 o'clock this morning, a statement by the President was issued at the White House that sixteen hours earlier—about the time that citizens on the Eastern seaboard were sitting down to their Sunday suppers—an American plane had dropped the single atomic bomb on the Japanese city of Hiroshima, an important army center.

Japanese Solemnly Warned

[illegible]

ROCKET SITE IS SEEN

125 B-29's Hit Japan's Toyokawa Naval Arsenal in Demolition Strike

[illegible]

ROOSEVELT AID CITED

Raiders Wrecked Norse Laboratory in Race for Key to Victory

[illegible]

MORRIS IS ACCUSED OF 'TAKING A WALK'

[illegible]

CHINESE WIN MORE OF 'INVASION COAST'

[illegible]

ATOM BOMBS MADE IN 3 HIDDEN 'CITIES'

Secrecy on Weapon So Great That Not Even Workers Knew of Their Product

[illegible]

TRAINS CANCELED IN STRICKEN AREA

[illegible]

War News Summarized

[illegible]

Turks Talk War if Russia Presses; Prefer Vain Battle to Surrender

[illegible]

Reich Exile Emerges as Heroine In Denial to Nazis of Atom's Secret

[illegible]

Figure 5.7. The front page of the *New York Times* on August 7, 1945, carried multiple stories with information about the Manhattan Project that had come from Laurence. (From The New York Times. © 1945 The New York Times Company. All rights reserved. Used under license.)

Laurence's letter to James some weeks earlier, the paper had heard nothing from its reporter since he had left on his secret assignment. The article did not quote from the letter to James—in fact, the newspaper never told its readers about the letter.[68]

As Catledge had decreed, the main story in the *Times* was written by Shalett. In all, ten of the paper's thirty-eight pages on August 7 were about the bombing, and much of that was based on Laurence's material.[69] On Tinian, Laurence did not immediately see the August 7 edition, but once he did, it produced a surreal reaction for him, recalling that "I had the curious experience of watching my own story in my own paper with a byline of somebody else."[70] The *Oak Ridge Journal* splayed Laurence's releases over more than four pages of the newspaper, but even in that venue, Laurence did not receive a byline. Instead, the stories were tagged "Official Release."[71] One scholar who examined initial coverage of the Hiroshima bombings by fourteen newspapers found that the press releases written by Laurence fostered public understanding of what had happened in Japan: "Because Laurence provided detailed information to reporters, most of the first stories carried fairly complete explanations of the history of the atom bomb project and the nature of nuclear fission."[72]

In his 1962 memoir, *Now It Can Be Told*, Groves recalled with pride, "Most newspapers published our releases in their entirety."[73] But privately he harbored second thoughts about his reliance on Laurence. If he could do it all over again, Groves later mused, he would have relied on several journalists, not just one. "They would have been the men who would have reached the greatest number of people in the country including press columnists and radio commentators," Groves wrote. "These men would have been shown enough so that when the break came, they would have been a very strong force for setting the proper course both in our internal problems and in our international ones." The journalists would have emphasized the importance of the army and "industrial management and developmental engineering," Groves wrote. His description was an implicit criticism of Laurence's approach, which strongly played up the role of scientists. Groves said that he wished that the press materials had featured "interviews and statements and articles by various distinguished figures, particularly in the scientific world, as well as in the industrial world instead of by the minor scientists." (Of course, Groves's own insistence on extreme compartmentalization meant that the only sources from whom

Laurence could have solicited comment were the "minor scientists" who were working on the project.)[74]

Nichols too approved of Laurence's work, but only because Laurence had obeyed orders. "He did a superior job, and I have never heard any implications that he violated secrecy. It was a fine example of military and press cooperation," Nichols wrote in his memoirs.[75]

After Hiroshima, Laurence kept busy waiting for the next bombing run. The next day Laurence flew on a reconnaissance mission over Hiroshima, but clouds obscured the bomb site.[76] Once the clouds cleared so a US Army Air Forces flight could photograph what remained of Hiroshima, he wrote a lengthy story that apparently never was distributed to the *Times* or any other newspaper. "The city of Hiroshima was practically wiped off the map in a manner more devastating than if it had been hit by an earthquake of the first magnitude," Laurence wrote. "The entire area within a radius of 10,000 feet from the heart of the city has been wiped clean as though it had never existed." Laurence estimated that 200,000 people lost their lives from the blast, but he did not once mention the prospect of radiation. The article also included after-action comments by those aboard the Enola Gay. "Everything just turned white in front of me," Laurence quoted Tibbets as saying of the blast. Laurence also described a Japanese-language leaflet "that will be dropped by the millions over Japan" urging surrender. "We are in possession of the most destructive explosive ever devised by man," Laurence quoted the leaflet as saying. "We have just begun to use this weapon against your homeland."[77]

On the evening of August 8, Laurence was sharing a beer with American scientist William G. Penny in the Tinian officers' club when he got word that he would fly on the next bombing run. A messenger told him, "Sir, you've got to get ready for the mission. It leaves tomorrow." They rocked the Quonset hut with their cheers. (Penny would go on the mission too.)[78] Missing the flight would have been "a sort of glorious journalistic strip tease, you might say—that I had seen everything but missed the dramatic event," Laurence later said, adding that his elation over joining the mission was mixed with a "certain tinge of regret" over the devastation that the bomb would cause.[79]

Once word came down that Laurence would be included on the second mission, the bombardier Thomas Ferebee helped Laurence prepare. "He took me to his locker on the airfield and he fitted me out with all the things

I needed," Laurence recalled. "First of all, I needed a parachute. And then you needed a survival belt, and then you needed an oxygen mask and a Mae West [life jacket]."[80] Years later, Tibbets admitted that his men overdid matters. "We hazed him a little bit by loading him up with all kinds of equipment he didn't need at all," Tibbets said. "We just loaded him up and finally wound up giving him a .45 caliber gun to shoot himself with in case they got captured."[81]

The crew shared the traditional preflight breakfast of ham and eggs, and the chaplain intoned a prayer.[82] Laurence rode to the plane in a jeep driven by Tibbets and boarded the instrument plane, which was to follow the main bomber and measure the force of the explosion.[83]

Ever dramatic, Laurence handed the radio operator his notebook. "Please take this notebook for me if we have to jump," Laurence asked. "Give it to the first American officer whom you see. I think you have a better chance of making it back to safety than I do." The operator replied, "Nothing will happen."[84] The plane was crowded, so Laurence perched on "a little instrument box," and he commenced writing as soon as the plane took off.[85]

But Laurence was confused about which plane he was aboard. The attack plane was flown by Charles Sweeney, who usually flew the Great Artiste. In his dispatch, Laurence wrote that the Great Artiste dropped the bomb on Nagasaki. However, for the Nagasaki mission, Sweeney and Fred Bock had swapped their usual planes, so the attack plane was Bockscar; Laurence—not the plutonium bomb—was aboard the Great Artiste.[86] Perhaps out of embarrassment, Laurence did not mention the plane's name in either *Dawn over Zero* or *Men and Atoms*. Years later, in an oral history, Laurence would cleverly use passive voice to paper over his error, saying, "At that time it was believed it was another plane, but it wasn't."[87]

As the planes flew through the night, Laurence ruminated about the personal destiny that had led him from Lithuania across the Atlantic Ocean to the United States and then into World War I, a career in journalism, and the encounter with atomic history. He said in an oral history that he excluded that in his Nagasaki story because he preferred to keep himself out of his writing. "I still haven't gotten used to seeing some of the newspaper reporters writing the big 'I,' with a capital letter, I, I, I. I have never used the term 'I' in any story I have ever written," he said at the time.[88] Yet his dispatch from Nagasaki was written in the first person, starting, "We are on our way to bomb the mainland of Japan," and he liberally sprinkled "I" throughout the dispatch. For example, the third paragraph began, "I watched the assembly of this man-made meteor during the past two days."[89]

He worried that the plane would crash, providing an opportunity for him to invoke his simulated rank after capture by the Japanese. "I said to myself, any minute now, you may become a colonel."[90] As the sun rose, the plane's radio operator asked Laurence if the bomb would end the war. "There is a very good chance this one may do the trick," the reporter replied. "If not then the next one or two surely will."[91]

Like the rest of the crew, he donned welding glasses as eye protection for the blast. "All of us became aware of a giant flash that broke through the dark barrier of our arc-welder's lenses and flooded our cabin with intense light," Laurence wrote. "We removed our glasses after the first flash, but the light still lingered on, a bluish-green light that illuminated the entire sky all around. A tremendous blast wave struck our ship and made it tremble from nose to tail. This was followed by four more blasts in rapid succession, each resounding like the boom of cannon fire hitting our plane from all directions."[92]

The bombers turned to return to Tinian. Running low on gas, the planes first landed in Okinawa to refuel. "Then we told them where we were, that we had dropped a bomb on Nagasaki, and, by God, we became the heroes of Okinawa right there. I mean the whole island came out to greet us," Laurence later recalled.[93] Note the first person "we"; Laurence certainly identified himself as part of the bombing crew rather than as an observer of it.

The planes touched down at Tinian at 10:30 p.m. to much less hoopla than had greeted the Hiroshima flight. "We were met by our ground crew and one photographer," Sweeney later recalled.[94] Nevertheless, the airmen partied through the night. Laurence took about ninety minutes to finalize his story.[95] "Laurence's 3,000-word story had clearance, but a military censor on Tinian made him boil it down to 500 words—and for some reason the dispatch was then shortstopped on Guam. It never got out at all," *Time* later reported.[96] After Laurence returned to Tinian after the Nagasaki bombing, Spaatz had him to lunch on August 9.[97]

After the two atomic bombing missions, the key participants held a press conference at Tinian. Laurence was there, although he was not quoted in any of the press dispatches. 1st Lt. Nicholas Del Genio, who had couriered uranium to Tinian, wanted to get the autographs of the participants. "I didn't have a piece of paper and didn't have time to get one," he recalled about two months later. "So I just pulled out a dollar bill and passed it down the line."[98] Laurence and ten others signed the front of the bill, and twenty-three others signed the back of the bill. The autographed dollar bill has been displayed at the American Museum of Science and Energy in Oak Ridge, Tennessee.[99]

In judging the ethics of Laurence's participation in the Manhattan Project, one must realize that he simply did not hold himself to today's journalistic standard of behavior. Rather, he conceived of himself as a science communicator trying to explain science and technology to a largely uninformed public in order to foster its acceptance. Of course, this unfolded in a wartime context in which the social goal was to win at virtually any cost. "Laurence's role as army staffer was no secret, as it has since been imagined," the historian Samuel Lebovic has written. "In the context of WWII press patriotism, it was a badge of pride."[100]

Other journalists also learned bits and pieces about the bomb project but suppressed the information before its use. Michael S. Sweeney has uncovered evidence that the liberal columnist Drew Pearson knew about the Manhattan Project and wrote nothing about it, other than a vague reference in a radio broadcast about two weeks after the Trinity test predicting a sudden victory over Japan.[101] Moreover, Laurence's coworkers at the *Times* apparently later fully understood and approved of Laurence's arrangement. Gelb later wrote, "While working with the Manhattan Project, Laurence was an employee of the government, not The Times."[102] And almost forty years after Trinity, a *Times* editorial invoked his example as evidence that the Pentagon should permit journalists to cover the US invasion of Grenada: "There's another necessity, the same one that led the Air Force to take William Laurence of The Times on the flight that dropped the atomic bomb on Nagasaki in 1945. Democracies depend on trust, and trust in war, small or large, depends on credible witnesses."

But some other journalists disapproved of Laurence's role, even at the time. Wilfred Burchett, who documented the horrors of Hiroshima shortly after the bombing, wrote acidly, "Few of his fellow journalists, much less his millions of readers, were aware of his real plenipotentiary status as the U.S. War department's nuclear propagandist."[103] Some critics focused on the tone and substance of Laurence's writing. "What seems so odd to a later generation is the fact that military censors passed—even, it seems, encouraged—Laurence's choice of tone," wrote Peter B. Hales, concluding that Laurence and his military overseers shared an attachment to "nature-worship, patriotism and religious righteousness."[104] Stewart K. Udall, the former congressman and secretary of the interior, called Laurence "the mythmaker-in-chief of the atomic age." Udall found particularly troublesome Laurence's preoccupation with drama, calling him "a frustrated dramatist who believed science had a sacred mission to 'save' civilization, and who saw an opportunity to embellish his eyewitness stories with millennial phrases that would make him the messiah of the new age."[105]

In recent decades, opinion has turned against Laurence. The science historian Alex Wellerstein sees Laurence as a clear ally of Groves: "Laurence represented himself as an independent journalist, but operated under the editorial finger of the Army Corps of Engineers, and used his position to push positions that were favorable to the government. His stories also neglected many aspects of the narrative that made Manhattan Project officials uncomfortable, like the discussion of civilian casualties or radiation injuries." Even one of Laurence's journalistic descendants at the *Times* has voiced disapproval. "This kind of relationship between a newspaper reporter and a government agency would be unacceptable today (though some people draw uncomfortable parallels with the reporters embedded with military forces in the war in Iraq)," Cornelia Dean, a former science editor at the *Times*, wrote in 2009.[106] In 2005 Amy Goodman and Juan Gonzalez tore into Laurence, calling him "a secret propaganda weapon" for the US military. "He was rewarded by being given a seat on the plane that dropped the bomb on Nagasaki, an experience that he described in the Times with religious awe."[107] Other journalists agreed. The *Virginian-Pilot* in Norfolk stated, "With benefit of hindsight, modern-day Americans should ask themselves: Is democracy served when journalists equate acquiescence to government policy with patriotism, as did Laurence?"[108] But some have defended Laurence. Media columnist Jack Shafer argued that "applying the standards of late-20th-century journalism to the newspapers of the 1940s . . . has only a limited relevance," because the *Times* shaped the media agenda much more than today and because walls between senior media executives and senior government officials were practically nonexistent.[109]

Back at Tinian after Nagasaki, the logical question for Laurence was what he would do next. Since Laurence was under the military's control, the immediate decision was not up to him. Once the bombings were over, the War Department shifted to removing American military staff from the theater even though Emperor Hirohito did not announce Japan's surrender until August 15. Word of Laurence's recall finally came on August 15 or 16, as Laurence sat inside a B-29 waiting to take off from Tinian for a reconnaissance mission over the two atomic bomb sites. "Five minutes later and I would have been off in the air," he later recalled. "That was one of my greatest disappointments, that I never got the chance to go to Japan to see the actual damage that was done over Hiroshima and Nagasaki."[110]

Even so, he didn't depart immediately; there were simply too many people to move. Laurence was among those named in an August 18 memo listing personnel who were to be relocated.[111] On August 23 Groves became insistent, instructing Farrell, "William L. Laurence should be returned to the United States without entry to Japan. He was employed because it was impossible for newspapermen to have access to the secrets of the project which would later be made public. That contingency no longer exists and there could be criticism if he were allowed to proceed as an employee of the War Department. Also the Times wishes his return."[112] On August 24 Farrell responded that "Laurence will get away as soon as his priority will permit him to get transportation or via a C-54 if one goes sooner."[113]

It took a little longer than Groves and Farrell wanted. Laurence landed back at Hamilton Field in California on September 2, his passenger information listing his occupation as "Gov't Official" and noting that he had surrendered a special passport that had been issued to him during his government service.[114] He was once again—officially, at least—an average citizen.

6

Aftermath

As the nation eagerly transitioned to the postwar era, Laurence received the welcome of a journalistic champion when he returned to the newsroom in New York. "He was a hero, and everyone applauded him," his friend Leslie Gelb recalled. "His manner, however, was modest. He carried himself with the quiet confidence of a man who knew he had nothing to prove."[1] Years later Laurence confessed that returning from the war was a letdown. "Everything from then on seemed that it might be an anti-climax," he recalled in an oral history.[2] He got no time off at all but was expected to resume his journalistic work immediately.

His first story after his return, on September 9, was a front-page eyewitness account of the Nagasaki bombing. He had written it right after the bombing, but the War Department had sat on the dispatch for weeks until finally releasing it to all newspapers for use on that date. Unlike Laurence's previous pieces for the Manhattan Project, his Tinian story was changed relatively little by the War Department. Most of the revisions seem to have been designed to protect operational details. For example,

Laurence's first sentence originally read, "We are on our way to bomb the mainland of Japan, in a formation equivalent to 2,000 and possibly 4,000, B-29 Superfortresses." But the publicly released version read only, "We are on our way to bomb the mainland of Japan." Just two sentences later, the approved release omitted information that the lead plane flew 5,000 feet ahead of its follower. A few paragraphs later the censor deleted Laurence's description of the plutonium weapon's casing as being "apple-green" in color.[3]

The approved version of the release also omitted a passage by Laurence emphasizing the bomb's safety; the deleted passage stated that the Nagasaki bomb could be "bounced around, dropped to the floor, or even shot at, without exploding." The editing of Laurence's dispatch also minimized any impression that Nagasaki had been chosen for bombing by chance. At a point in the narrative at which the bomber is waiting to determine its target, a deleted passage stated, "But at this point no one knows which of these cities will be chosen for annihilation. The final choice lies with destiny. The winds over Japan will make the decision. If they carry heavy clouds over our primary target that city will be spared, at least for the time being. None of its inhabitants will ever know that a wind of benevolent destiny had passed over their heads. But that same wind will doom another city, our secondary target." Before the target was set as Nagasaki, Laurence wrote, two of the bombers had circled Yakushima while waiting for the third bomber to arrive. The War Department deleted the island's name, leaving the article to state that the bombers had been circling without saying where. The War Department did not correct one key error by Laurence: He confused the names of the warplanes used in the Nagasaki attack and misstated that the bomb was dropped from the Great Artiste, not Bockscar as had been the case.

In the copy of Laurence's story distributed to newspapers, the War Department added the note, "The following release was written by William L. Laurence, Science writer for the New York Times, and Special Consultant to the Manhattan Engineer District and former Pulitzer Prize winner. The story can be released with or without the use of Mr. Laurence's name."[4] The *Times*, of course, did use Laurence's name above the story, which it ran on its front page. Above his byline, the paper ran an editor's note: "Mr. Laurence, science writer for The New York Times and a Pulitzer Prize winner, is a special consultant to the Manhattan Engineer

District, the War Department's special service that developed the atomic bomb."[5] The version of the story published in the *Times* was identical to the press release aside from minor stylistic changes.

The Associated Press did not transmit Laurence's piece verbatim but instead distributed a story that quoted from it extensively. Many newspapers across the country ran this story on the front page; other newspapers, such as the *Los Angeles Times*, reprinted Laurence's dispatch. Although the atom bomb was a local story in Tennessee, the *Knoxville Journal* still ran Laurence's story on page 3, perhaps reflecting some fatigue about atomic news only one month after their historic battlefield use.[6] The *Oakland Tribune* compromised by running the AP story on page 1 and Laurence's dispatch on page 2.[7] Coverage extended overseas; the *Sunday Times* in London ran a Reuters version on the front page, and it also appeared on the front page of a Scotland newspaper.[8] Australian newspapers ran an Australian Associated Press story, often on the front page.[9]

In New York the popular mayor, Fiorello LaGuardia, heaped praise on Laurence's article during a weekly radio address to the city. "Have you read the article by William L. Laurence in today's Times?" he asked listeners. "I recommend that to every current events class in the high schools and colleges. That article can be read and reread and when it becomes stale as a subject for current events, passed on to the Department of Literature. It is a classic in the English language and will remain among the English classics."[10]

But relief over the US victory in Japan already was giving way to gnawing concerns, in America and overseas, about the possibility of long-lived and highly poisonous radioactivity that the atomic bombs may have deposited at their bombing sites. This directly contradicted the US government's position that the atomic bomb was deadly only because of its enormous blast, extreme heat, and powerful but short-lived radiation released in the blast; the government held that radiation no longer would be a problem after the blast subsided. But on-the-spot reports from Japan held otherwise: Wilfred G. Burchett, an Australian journalist who was the first non-Japanese journalist to visit Hiroshima after the bombing, produced powerful reporting on the plight of the survivors, emphasizing the continuing radiation. The *Times* added to those concerns when William H. Lawrence—"Political Bill" as *Times* insiders called him to avoid confusion with Laurence—visited Hiroshima and reported on September 5 that survivors of the atom bomb were dying at the rate of one hundred each day.[11] The political damage from that report was not undone when, on September 13, the *Times* ran another report from Lawrence quoting Gen. Thomas Farrell, who had visited the bomb site, claiming that the atom bomb had left no persistent radiation at Hiroshima. The bomb primarily killed through its blast, not radiation, Farrell insisted.[12]

The September 5 Lawrence article produced enduring confusion about William Laurence's role in covering the early controversy about the radiation effects of the two atom bombings. Even at the time, some reporters confused Laurence and Lawrence. Reporting on Lawrence's September 5 story on his Hiroshima visit, the *Times* of London incorrectly credited the report to "Mr. William L. Lawrence [*sic*] of the New York Times science staff."[13] Over the years, many historians and critics have misunderstood the September article about Hiroshima to have been written by Atomic

Figure 6.1. William L. Laurence (left) was sometimes confused with William H. Lawrence (right), a political reporter for the *Times*. Other staff members took to referring to Laurence as "Atomic Bill" and Lawrence as "Political Bill." (New York Public Library)

Bill rather than Political Bill. Burchett conflated Lawrence and Laurence in his 1983 book, *Shadows of Hiroshima*, which was a fuller presentation of Burchett's 1945 reporting trip to the remains of the bombed city. Burchett spent several pages excoriating Laurence's supposed reporting from Hiroshima—reporting that never happened. "I reported what I had seen and heard, while Laurence sent back a prefabricated report reflecting the 'official line,'" Burchett wrote thirty-eight years after the bombing.[14] Although Laurence unquestionably downplayed the issue of radiation at other times, this particular incident is not validly charged against him.

Even so, Laurence—along with other journalists—is justifiably blamed for participating in another journalistic escapade that also downplayed the risk from radiation from the bomb that played out only a few days later. Lt. Gen. Leslie R. Groves, the head of the Manhattan Project, invited a select group of journalists to the Trinity Site in mid-September to see for themselves that the concerns about radiation at Hiroshima and Nagasaki were overblown. Laurence was an eager participant.

Photos of the event show Groves, Oppenheimer, Laurence, and other visitors gathered around the ruin of the tower that previously had held the Gadget, all looking a bit incongruous in jackets and ties. Tellingly, the visitors wore white coverings on their shoes to prevent them from tracking radioactive material from the ground. *Life* magazine ran a full-page picture of the surface of the crater from the Trinity explosion, covered with a greenish glass formed when the blast's heat fused sand on the desert floor; another picture in the magazine depicted Laurence standing atop a lead-lined Sherman tank that had been used by the Manhattan Project scientists to shield themselves from radiation.[15]

In his article for the *Times* on the Trinity show-and-tell, Laurence asserted that the visit "gave the most effective answer today to Japanese propaganda that radiations were responsible for deaths even after the day of the explosion, Aug. 6, and that persons entering Hiroshima had contracted mysterious maladies due to persistent radioactivity." Laurence wrote that most of the deaths had occurred from the force and heat of the blast and the fires that the blast triggered. As Laurence and other reporters walked around the Trinity Site, Geiger counters "showed that less than two months after the explosion the radiations on the surface had dwindled to a minute quantity, safe for continuous habitation."[16]

Figure 6.2. A press tour of the Trinity Site in September 1945 was intended to reassure the public about the absence of lingering radiation from the atomic test there two months previously. The physicist Robert Oppenheimer and army Lt. Gen. Leslie R. Groves examined what remained of the steel tower that had held the Gadget. (Library of Congress, Prints and Photographs Division, NYWT&S Collection, LC-USZ62-128768)

Groves appreciated the article. Within hours of its publication, he wrote Sulzberger to thank him for Laurence's participation in the Manhattan Project. "I want you to know that he has done splendidly," Groves wrote. "He is a master craftsman in his field and the leading journalistic expert in the United States on atomic energy. I am sure that he is a credit to your organization as he was to ours."[17] Two weeks later, Sulzberger wrote back to return the compliment. "Bill Laurence did a fine job—of course, we're proud of him," Sulzberger responded. "That feeling isn't restricted to members of our staff—we're proud of you too, Sir!"[18]

Figure 6.3. Laurence and Groves at the Trinity press event in September 1945. Laurence's story about the event in the *Times* repeated the government's position that the test showed that radiation did not persist after the explosion. (Atomic Heritage Foundation)

The *Times* then switched into overdrive in publishing articles by Laurence using his reporting from the Manhattan Project. From September 26 through October 9, the *Times* published ten stories by Laurence summarizing what he had seen and learned. Perhaps suggesting that his editors already were suffering atomic fatigue, only the first story in the series—again describing the Trinity test—started on the front page. The others started inside, with one of them buried as deep as page 10.[19] The *Times* did not use this as an opportunity to sneak classified information into print, James reassured Groves: "I have told him to write nothing without clearing it with the War Department."[20]

The *Times* did its best to leverage the fame of its atomic reporter, adorning its delivery trucks with placards proclaiming, "Read the eyewitness story of the atomic bomb by William L. Laurence." He also appeared on the

Times's own radio station, WQXR, on November 20.[21] But if the *Times* was building its reputation and circulation on Laurence, Laurence himself wasn't seeing any return. Laurence had expected to get at least a $100 weekly raise when he returned from the Manhattan Project but received nothing. "Months passed and I got no recognition whatever. They didn't give me a dime increase," he recalled two decades later. After six months, Laurence raised a stink, and his pay increased by $25, to $256 weekly.[22]

After publishing this package of Manhattan Project stories, Laurence took an extended break from writing for the *Times*. With the exception of one story on November 17, 1945, Laurence's byline did not appear in the *Times* between October 9, 1945, and March 12, 1946. When he resumed publishing in the paper, he wrote predominantly on topics relating to the atom bomb and world politics rather than the wide array of scientific topics that had occupied him before the bomb. Of the forty-six stories that the *Times* published with his byline in 1946, twenty-seven were on topics related to atomic energy. In all of 1945, the *Times* published only sixteen articles bearing his name, fewer than in any year since 1932.

Although he wasn't writing for the paper, Laurence was still busy—becoming a personality. His firsthand knowledge of the mysterious and terrifying atomic bomb, and his willingness to talk about it in the most dramatic terms, made him a popular speaker and an intriguing public figure. The *New Yorker* rushed a profile of Laurence into print in its August 18 issue, containing up-close-and-personal details such as his Dachshund named Einstein and his apparent obsession with rivers that led him to place mirrors throughout his apartment to afford frequent easy views of the East River.[23] On October 10, 1945, New York City radio station WNEW broadcast a fifteen-minute radio drama relating Laurence's coverage of the discovery of fission in 1939 and the Manhattan Project. The show cast Laurence as a visionary reporter who told his managing editor, Edwin James, about the bomb long before the Manhattan Project was made public, saying, "I want to study every angle of this atomic energy affair, and try to awaken people here to its potentialities." The radio script had James responding, "Well, what are you waiting for? Get to work!"[24]

Laurence also volunteered for a federal government project to raise money from the public called the Victory Loan drive, the eighth such fundraising effort since the start of World War II. Laurence participated in the drive's Book and Author Rally, in which a number of literary luminaries spoke about their ideas and raised patriotic themes in order to stoke public support for the fundraising effort. Laurence had purchased war bonds as a soldier during World War I, so the drive appealed to his patriotism. Appearing on a platform with other prominent writers and intellectuals also afforded Laurence a venue for promoting his ideas regarding atomic bombs and in general increasing his reputation.

Nine other authors also were on the tour: Clifton Fadiman, Bill Mauldin, Edna Ferber, Carl Sandburg, Lillian Hellman, Carl Van Doren, Bennett Cerf, Fanny Hurst, and Louis Bromfield. During the late fall of 1945, they visited Gary, Indiana; Milwaukee; Buffalo; Williamsport, Pennsylvania; Harrisburg; Nashville; Atlanta; Miami; and Winter Park, Florida. They raised more than $14 million. Laurence was the highlight, not surprisingly since the Hiroshima and Nagasaki bombings had happened less than three months earlier and Laurence was widely seen as having exclusive knowledge of the still-mysterious technology and its implications for society.[25]

One such rally happened on November 2, a breakneck day of appearances in Cleveland, starting with a breakfast with local reporters and moving on to talks at a local employer and a local high school, a book signing at a bookstore, a discussion broadcast on a Cleveland radio station, and a visit to a local hospital. "Nobody blamed Mr. Laurence much for not being able to get going early enough for the first event of the day," a Cleveland newspaper reported sympathetically.[26] The next day, Laurence and the contingent addressed 3,500 people in an auditorium in Gary. The program opened with several musical numbers. Then Clifton Fadiman, the moderator of the popular *Information Please* radio quiz show, introduced Laurence, who told the audience, "The bombs exploded over Hiroshima and Nagasaki puts [*sic*] us all under sentence of death. That death will come if we fail to recognize the implications of the atomic bomb, which has within it the possibility of destroying a large percentage of the world's population." The Gary newspaper said that Laurence "held the audience spellbound with his delineation of the majestic beauty and breath-taking power of the experimental bomb explosion in New Mexico, which he witnessed from a distance of five miles, and of the even greater phenomenon which rose to a height of 60,000 feet over Nagasaki." The other speakers—Ferber,

Mauldin, and Sandburg—"made only brief appearances" due to the length of Laurence's comments.[27]

Laurence also appeared on the radio. On October 8, 1945, he was a guest on *Information Please*, where experts tried to answer questions submitted by listeners. And on the evening of November 16, 1945, Laurence appeared on Mutual's *Meet the Press* radio program, forerunner of the television show of the same name on NBC. He spoke about the risk of atom bombs making Earth uninhabitable, the need for more nuclear tests, and outlawing war.[28]

The speechmaking circuit also attracted Laurence. In November 1945 he told one audience that the use of the atom bomb by the United States had saved as many as 1 million lives.[29] In December he told the annual convention of the National Association of Manufacturers that "if General Groves were given full authority, such as he had on the original development of the bombs, we could be using atomic energy for peaceful purposes in less time than it took us to build the first bomb."[30]

A key theme in Laurence's speeches was his endorsement of international control of atomic power. On the evening of November 28, 1945, Laurence addressed Americans United for World Organization, a group that favored world government, in the grand ballroom of the Waldorf Astoria hotel in Manhattan. "There can be only one protection against the atomic bomb—PEACE," Laurence told the audience. "Only the peoples of the world can supply peace. The question we face cannot be answered by Americans alone. No unilateral decision will suffice."[31]

But Laurence soon found that publicity can be a double-edged sword. The *New Yorker*, which had helped trigger the growth of Laurence's public profile soon after the atomic bombings, instead skewered him months later, in November 1945. Frank Sullivan, the American humorist, poked fun at Laurence in Sullivan's long-running series of articles in the magazine featuring the fictitious Mr. Arbuthnot, a mangler of clichés. In "The Cliché Expert Testifies on the Atom," published in the *New Yorker* three months after the bombings, Mr. Arbuthnot's interlocutor asks for an explanation of atomic energy. "Well, listen carefully and I'll give you a highly technical explanation," Mr. Arbuthnot replies, and then spews out a litany of jargon concluding, "heavy water . . . New Mexico . . . mushroom-shaped cloud . . . awesome sight . . . fission . . . William L. Laurence . . . and there you had a weapon potentially destructive beyond the wildest nightmares of science. Do I make myself clear?"[32]

During his break in writing for the newspaper's daily edition, Laurence also worked on writing *Dawn over Zero*, a book based on his Manhattan Project experiences. His Manhattan Project work helped him connect with a publisher: in its first issue after the atomic bombings of Japan, the military magazine *Army and Navy Journal* had run Laurence's press release about the Trinity test without a byline. The article caught the eye of the book publisher Alfred A. Knopf, who wrote to the magazine to ask who had written the article, which he declared to be "one of the best things of the kind I have ever read." Knopf said in the letter that he was interested in asking the author to write a book.[33] The magazine's editor identified Laurence as the author. But even before the editor replied, Knopf had reached out to Laurence on the strength of the August 18 *New Yorker* profile of Laurence. The two had met before, and Knopf wrote, "It is a long time since we have met and I begin to understand at least one reason why. But I haven't lost my old interest in the possibility of a book from you."[34] By September 6—one month after the bombing of Hiroshima—they had come to an informal agreement to pay Laurence an advance of $5,000 for a book on "the Atomic Bomb."[35] By December 1945 Laurence's agent began pressuring Knopf to publish the book in spring 1946, so that it would be available in bookstores in May 1946 if Laurence were awarded a Pulitzer Prize that spring.

However, the manuscript landed with a thud at Knopf's offices in February 1946: the publisher's reviewers found it repetitive. One said she found "nothing new and that she did not think it worth while" to ask any additional referees to read the manuscript. When a Knopf staffer called Florence Laurence to discuss the reviews, because her husband was traveling, she was "on the truculent side, convinced that her husband had produced a masterpiece," the staffer reported. An executive at the publisher wrote to Laurence's agent to prepare them for dealing with Mrs. Laurence, warning that the manuscript was filled with repetitive passages and poorly organized into newspaper-article–sized chunks. The executive added: "A great deal of the writing was frightfully bad."[36] But in March Knopf paid the $5,000 advance against future royalties from book sales and proceeded with publication.[37]

When Laurence wasn't working on writing his book, he was writing for publications other than the *Times*. He wrote a lengthy article on the development of the atom bomb for the first issue of the annual *Information Please Almanac*. It championed the prospects for civilian use of nuclear power for generating electricity and for medical treatments but sought to debunk some of the more fabulous visions of nuclear power, such as the notion that a nuclear power plant could be easily miniaturized.[38] A reviewer for the *Times* wrote that the almanac's articles were "sprightly done, notably so in the piece on the evolution of the atomic age, as carefully traced by its Boswell, William L. Laurence."[39] The *Times* even used a passage from Laurence's almanac article as the solution to a word puzzle.[40]

Laurence also returned to a fascination with biomedical research that, in his view, could reverse or avoid aging and death. In early 1943, before Laurence's Manhattan Project assignment, American journalists had begun writing about Alexander Bogomolets, a Ukrainian pathophysiologist who developed a biomedical treatment that he claimed extended human life. When being recruited to work on the Manhattan Project, Laurence had agreed but asked for two weeks to tie up some loose ends. One of those loose ends was finishing up an article on Bogomolets for *Ladies' Home Journal*, which the magazine nonetheless did not publish until after the war. Laurence wrote that Bogomolets's research had created excitement unseen since the discovery of penicillin.[41] A condensed version of Laurence's article subsequently appeared in *Reader's Digest*.[42] It also was reproduced in an annual anthology of the year's most influential science news stories, a series that had not included his prior articles on the atom-bomb coverage.[43] Some researchers were appalled by the attention that Laurence lavished on Bogomolets. The *Journal of Gerontology* editorialized, "It is unfortunate that Mr. Laurence has used his great gifts of lucid exposition to promise quick clinical relief from the infirmities of old age, including arteriosclerosis, hypertension, and cancer as he did in this article. The most enthusiastic investigator or physician would hesitate to issue a dated promissory note as does Laurence when he says that 'the Bogomolets anti-age serum can be made available to the American people (some two or three years hence).' "[44]

Laurence never wrote about Bogomolets for the *Times*. That fell to Kaempffert, who was as skeptical about Bogomolets as he was of many of

Laurence's other pet scientific topics. In May 1946 Kaempffert wrote that in one test by Bogomolets on forty-eight patients, "the results were none too good," with slight or definite improvement observed in fourteen of the forty-eight.[45] Two weeks later, Bogomolets gave an interview to American and British reporters in Kiev, including a *Times* correspondent, in which he said that his serum could enable humans to live to age 150. He rebutted Western criticisms of his research, saying that he never had stated that his serum was a cure for cancer.[46] The next day, in an editorial that was unbylined but most likely was written by Kaempffert, the *Times* said that Bogomolets's claims were "much more temperate than that of his American interpreters. . . . Bogomolets himself admits that good food, exercise, a sane way of living cannot be ignored even if his serum is taken."[47] About a week later, Kaempffert reviewed Bogomolets's book and took a perplexed stance, writing, "Has Bogomolets made an important discovery? Probably. But it is hard to evaluate it." Still, Kaempffert concluded that his position that humans can live to 150 is "not unreasonable."[48]

As on so many other topics, in retrospect Laurence's optimism about Bogomolets was ill founded. Bogomolets's treatment "never entered the mainstream of immunology and medicine," one history of immunological research states. Bogomolets's claims were vague and difficult to verify. "Its early widespread clinical application in the Soviet Union seems to be yet another manifestation of the uncritical pursuit of biomedical science in that period of the Soviet Union's history." Few papers were published in respected Western medical journals. Four different American researchers found little or no benefit from the serum. "Interest in it faded away in the early 1960s."[49]

But Laurence's writing on Bogomolets did have at least one unexpected, long-lasting impact on the field of gerontology, by inspiring a researcher to enter the field. Denham Harman was a budding chemist who studied at the University of California, Berkeley and eventually earned a doctorate while working at Shell Oil in Emeryville, California. His research came to center on the use of free radicals in manufacturing chemicals. Laurence's article on Bogomolets inspired him to change careers and attend medical school. Harman went on to develop the free-radical theory of aging, which holds that highly reactive chemicals in the body cause aging by damaging cellular structures; the theory, which has become popular, maintains that aging can be retarded by the use of antioxidants.[50]

Still, Laurence's fame centered on his writing on the atom bomb. In late January 1946 the *Times* nominated Laurence for his second Pulitzer Prize in the category of "a distinguished example of a reporter's work during the year, the test being accuracy and terseness." The nomination package included his *Times* articles on the Manhattan Project published in September and October 1945. The paper's nomination statement did not note his association with the government, but an introduction published atop his article on the Trinity test, submitted as part of the package, did state that Laurence "was detached for service with the War Department at its request to explain the atomic bomb to the lay public."[51] Columbia University's journalism school, which administers the Pulitzer Prizes, received forty nominations for the reporting award. Paul Friggens, a Columbia staff member, wrote the university's acting president to summarize their conclusions about the nominees: "I must unquestionally [*sic*] urge for a Pulitzer Award the atom bomb work of William L. Laurence of the New York Times. This award should be based on his eye-witness account of the bombing of Nagasaki and his sebsequent [*sic*] ten articles on the development, production and significance of the bomb."[52]

Crucially for criticism of Laurence's selection that arose decades later, the memo from Friggens explicitly acknowledged Laurence's connection to the government while he was reporting. He wrote, "At the same time, I should like to point out to the Committee that Mr. Laurence was assigned to this work by the War Department from the beginning, that he alone among reporters was permitted to study the bomb, to witness the first trial in New Mexico and to see one of the bombs dropped on Nagasaki."[53]

Columbia announced Laurence's selection for the Pulitzer on May 7, 1946. The *Times* ran the announcement on the front page but did not mention Laurence in the headline. Rather, his name appeared only in the body of the article after two playwrights who shared a Pulitzer for playwriting and *Times* colleague Arnaldo Cortesi, who had won for foreign reporting.[54] Three weeks later, Laurence's alma mater, Boston University, awarded him an honorary Doctor of Science degree.[55]

Groves wrote Laurence and James to congratulate them. He told Laurence, "It was no real surprise as I was certain it would be coming your way as soon as I saw the article for the first time." To James, he recalled his vague request for Laurence's services: "I know that you must feel pleased over your perspicuity in acceding to my request for the loan of Mr. Laurence to our work, particularly when I recall how meager and nebulous was the story I told you at that time."[56] Laurence responded to Groves with an effusive letter of his own. "I shall never forget your kindnesses to me during the never-to-be-forgotten days in which I had the great privilege of serving under you," he wrote, "assuring you of my deep devotion and loyalty to you personally and of my high regard for you as one [of] our greatest living Americans."[57]

Although Laurence had no hesitation touting his two Pulitzer awards in subsequent years, the accomplishment left him unfulfilled. After receiving his second Pulitzer, Laurence told a *Times* colleague, "When I first came here to this country . . . I wanted to be a great writer. . . . I never wanted to be a reporter, really. Winning Pulitzer Prizes is still not the same."[58]

7

Atomic Plagiarism in the South Pacific

How dramatically things can change in one short year. When Laurence covered the first atomic explosion at Alamogordo in July 1945, he was the only reporter on hand, huddling for safety with a handful of scientists in the predawn darkness of the New Mexico desert. A year later, at the fourth atomic blast, the so-called Able experiment of the War Department's Operation Crossroads, Laurence shared the story with almost two hundred other journalists, decked out in tropical shirts and sandals, watching one of history's biggest demonstrations of military power from the deck of a navy destroyer in the South Pacific.

Operation Crossroads was the US government's first set of nuclear tests after World War II and the kickoff of many more that the United States would set off in the South Pacific over the ensuing years, converting the lovely but so-called developing region into a nuclear wasteland and forcing the permanent relocation of many of its residents. In Crossroads, the US military arrayed ninety-five unstaffed warships—mostly American, but captured Japanese and German vessels as well—around Bikini Lagoon to see what a nuclear explosion would do to them. In the first Crossroads test, code-named Able, a plutonium weapon of the same design used at Nagasaki was dropped by plane on the ghost fleet. The second test, code-named Baker, involved detonating a similar weapon ninety feet underwater in the lagoon. A third test, named Charlie, was to be exploded even deeper underwater outside the lagoon, but that test never took place.

Crossroads was the largest peacetime US military exercise to that date, involving 45,400 men, 220 ships, and 160 aircraft. The stated objective was to evaluate the effects of nuclear blasts on naval ships and equipment. Others saw a different motive: to emphasize the atomic strength of the US military. Unlike the secrecy that had shrouded the Trinity test, the government announced Crossroads months in advance, and journalists and observers from around the world were invited to observe the tests and their aftermath.

Although the US Army's Manhattan Engineer District was still in existence, the Joint Chiefs of Staff assigned responsibility for Operation Crossroads to the newly created Army-Navy Joint Task Force One, under Admiral William H. P. Blandy. The government tried to recruit Laurence to work on publicity for Crossroads, but he suggested they approach Walter Sullivan, a junior reporter (and future science editor of the *Times*). Sullivan also declined because he was expecting a new post at the *Times* as a foreign correspondent.[1]

Journalists understood that Crossroads was a news event not to be missed, but many did because space was so limited. Arthur Hays Sulzberger, publisher of the *Times*, was one of the 3,500 journalists who applied for permission to cover the test. Like the overwhelming majority of applicants, Sulzberger was turned down. Laurence did receive accreditation, as did Hanson W. Baldwin, the military editor of the *Times*.[2] Laurence and the other accredited journalists were instructed to report to the press ship, the USS *Appalachian,* in Oakland, California, on June 12, or at Honolulu on June 19.[3] The navy would transport them to the South Pacific and back.

The navy assigned rooms in the *Appalachian* by reporters' ages. The oldest accredited journalist, the longtime war correspondent Frederick Palmer, drew a private room that featured a desk, locker, and fan. Laurence, at age fifty-eight, was the fourth oldest of the correspondents, and he received a private room too.[4] Robert D. Potter, who covered Crossroads for the Hearst Sunday newspaper supplement *American Weekly*, recalled that Laurence and his fellow science journalists at Crossroads "griped about many things, but not about the assignment of space on the *Appalachian*."[5]

The Crossroads project was controversial among some scientists. In May 1946, two months before the Crossroads tests, Lee A. DuBridge (a physicist who had headed the national radar program at MIT during the war and who went on to become president of the California Institute of Technology from 1946 to 1969 and science adviser to Presidents Truman, Eisenhower, and Nixon) criticized Crossroads as having "no military or scientific value." DuBridge maintained that the atomic bomb was an impractical weapon against ships, and he asserted that the Crossroads tests were arranged in such a way that its explosions were unlikely to sink many ships.[6] DuBridge expanded the audience for his argument by sending a letter to the editor in the *Times* in early May. "I think it is safe to dismiss as negligible the scientific value of the tests," he wrote in the newspaper, adding a prediction that "only three or four out of ninety ships will be damaged."[7] However, such objections never appeared in Laurence's reporting. After an April press conference by Crossroads officials, Laurence wrote a lengthy front-page story describing the project's scientific objectives and instruments. Laurence characterized Crossroads as "the most stupendous single set of experiments in history." He mentioned no criticisms of the project.[8]

If DuBridge argued that Crossroads was a waste of time, Anatol J. Shneiderov of Johns Hopkins University warned that the project could prove downright cataclysmic, if the blast cracked the planet's crust. "The author is afraid that there may be no survivors to report on the experiment if the bomb of Hiroshima-size is exploded on or under the ocean surface," Shneiderov wrote in the May 31 issue of the journal *Science*. That wasn't even the scariest part of his prediction; Shneiderov further warned that seawater spilling into the crack could generate both a tidal wave that would overpower ships and a cloud of dust and steam that could block the sun for decades.[9] Laurence and the *Times* ignored Shneiderov's forecast, but the Associated Press's Howard W. Blakeslee reported it on June 10, two days before the *Appalachian* was scheduled to leave Oakland, with San Francisco newspapers picking up the dispatch.[10] The *San Francisco Examiner* headlined Blakeslee's article "All A-Bomb Personnel May Die, Scientist Says."[11] The scaremongering news coverage panicked both sailors who were getting ready to set sail and their parents, some of whom cabled the ship's executive officer demanding that their sons be discharged from the ship in Honolulu.[12] The *Appalachian*'s executive officer insisted that Blakeslee help settle the crew. Blakeslee used the public address system of the three Crossroads ships still in port—the *Appalachian*, the *Panamint*, and the *Blue Ridge*—to reassure their crews that other researchers had discounted Shneiderov's conclusions, an assertion that he had not included in his article.[13]

The *Appalachian* left Oakland on June 12, carrying Laurence and virtually all of the other journalists. Scientists and officials were located on other ships, giving navy officials extensive control over the information flowing to the reporters. This eliminated the need for heavy-handed censorship but also created a public stage for gregarious, knowledgeable reporters like Laurence who were eager to hold forth for the other journalists. Potter recalled, "The correspondents could write anything they wanted but their sources of information were practically zero—especially on the way to Bikini. Thus it was that NASW science writers like Dave Dietz and Howard Blakeslee, along with Laurence and Potter, were the oracles of the moment. At least they knew the difference between fission and fear, between a neutron and nonsense and between energy and MC squared."[14] Historian Jonathan Weisgall concluded that Laurence emerged as "the undisputed dean of correspondents covering Crossroads."[15]

Other journalists at the time similarly characterized Laurence as an oracle—a metaphor that emphasized both the mysteriousness of nuclear weaponry and Laurence's privileged access into the heart of that mystery. Laurence conducted informal tutorials for journalists on the *Appalachian*. "Being the only man in the world who has done that [witnessed Trinity and Nagasaki], he's something of a sea-going oracle," one reporter wrote.[16] The trade journal *Editor and Publisher* wrote that Laurence was "the oracle to most other correspondents, being the only one present to view the New Mexico test."[17] Yet another reporter referred to him simply as the "venerated William Laurence."[18] A correspondent for the *Denver Post* wrote, "Short, bespectacled but unprofessorish Laurence, with his shock of graying hair, occupies a unique position among group correspondents aboard Appalachian, probably as inclusive a gathering of journalistic big names as gathered on single story this or any year. His intimate knowledge of Manhattan district before and after Hiroshima, plus [the] fact he saw initial New Mexico bomb test then flew aboard bomber which devastated Nagasaki, makes him unfailing source first-hand information to other correspondents."[19]

But Laurence didn't just coach and instruct the Crossroads journalists; he also allowed them to cite him as an expert. The city editor of a Washington state newspaper quoted Laurence as assuring him that the federal government's plutonium factory in that state had a secure future because plutonium would play a crucial role in both military and civilian uses of atomic power.[20] On June 14, with the press ship between Oakland and Honolulu, Bernard Baruch announced the US government's proposal for an international atomic development authority that would oversee nations' use of atomic power as a means to preventing atomic proliferation.[21] The journalists on the ship turned to the only atomic experts they had at hand: Laurence and the few other seasoned science journalists. Laurence immediately endorsed the Baruch plan, asking, "If you don't like this plan, what else is there you can propose?" He saw no alternative. "Either this or a war with atomic bombs!"[22] In one journalists' roundtable, Laurence called the Baruch report the "greatest document in modern history. It may well change the course of the world. It's the first time in history that any nation has voluntarily offered to surrender any of its sovereignty."[23]

Laurence's support for the Baruch plan echoed around the globe. A reporter from the Australian Associated Press filed a story, published in several newspapers across Australia, in which Laurence was quoted as calling the plan a "second Magna Carta." Laurence said: "It puts any nation which raises difficulties to international control, or seeks to cling to the power of veto, squarely in the position of a potential aggressor."[24]

Long after returning to the mainland after Crossroads, Laurence would remain an advocate for the Baruch plan. He championed it for months afterward in speeches to civic and school groups, such as a December 1946 appearance at Trinity College in Hartford, Connecticut, where Godfrey Nelson, a *Times* attorney, had a son and arranged for Laurence's appearance. "There is and can be no other plan," Laurence told the students. "There can be no possible compromise. It is therefore the duty of the American people to get acquainted with the basic facts of the plan, and to get behind it with the weight of public opinion."[25] Laurence spoke in favor of the Baruch plan in Iowa and New York too.[26] *Times* editors were very well aware of Laurence's activism on the issue. In September 1946 the paper ran a brief story covering a speech by Laurence to the Overseas Press Club in which he strongly endorsed the Baruch plan.[27] *Times* executives sought to leverage Laurence's prominence on the issue for the paper's benefit, by having him speak on the issue at a company event for Detroit executives later that year.[28] The Baruch plan died in 1948 due to opposition from the Soviet Union.

As always, Laurence's presence was news. When the *Appalachian* arrived in Hawaii on June 18, the *Honolulu Star-Bulletin* noted that the reporters on board the ship included "science writers, experts in their field who write learnedly on subjects beyond the understanding of less educated minds. Perhaps the best known of this little group of specialists are Howard Blakeslee, Associated Press science editor, and William Laurence of the *New York Times.*" Laurence told the reporter, "I am not here as a science writer. I have written all the science. I am here as a reporter to write about a spectacle."[29]

The *Appalachian* left Hawaii for Bikini on June 20. On that leg of the journey, news was scarce. Some of the reporters were so bored that they resorted to writing about Gladys, a wooden penguin toy owned by one of the reporters. Set atop a glass of water, it continually dipped its beak in the water, rocking back and forth. Perplexed journalists turned to Laurence for an explanation. One reporter quoted Laurence as saying, "I'm a reporter, and I hope a good one. I can report what Gladys does. But why? I don't know. It's beyond me."[30] Another *Appalachian* journalist was reduced to polling the reporters for their predictions about whether Joe Louis or Billy Conn would win a much-anticipated boxing match on June 19, eleven days before the expected date of the Able blast. Laurence predicted that Louis would win in the third round, while Louis actually knocked Conn out in the eighth round.[31]

Blakeslee and Laurence continued to labor to contain the apocalyptic predictions about the Crossroads tests. "Dr. Dubridge [*sic*] is not speaking as a scientist when he criticizes the test before it has taken place," Laurence told journalists. "When scientists say such a test will be useless before it has been conducted they are speaking as political propagandists," he said. He also criticized Shneiderov as "a cheap attempt at scare journalism, unworthy of a scientist."[32]

Meanwhile, the military's briefings for journalists on substantive topics continued, including one on June 21 that featured Laurence, who told reporters they should expect the Able blast to produce a mushroom-shaped cloud and warned them to use the special protective glasses that they would be issued. "Don't fail to put them on and don't look directly at the brilliant pin point or you may miss the real show, which you have traveled thousands of miles to see."[33] Laurence left some journalists uneasy. A Canadian journalist observed, "There are some aboard this vessel who already have misgivings about their assignments to this expedition and Laurence's vivid recital didn't make them feel any easier in mind."[34]

On June 23 some of the journalists composed a song satirizing Laurence, Blakeslee, and navy officers, sung to the tune of "When I Was a Lad" from Gilbert and Sullivan's *HMS Pinafore*. At least three journalists filed stories on the song, including Blakeslee. The lyrics varied slightly in different reports. Blakeslee's version included the following lines:

> Blakeslee and Laurence just sit and think
> But drinkers have good cause to scoff
> For the gol-blamed thing may never go off.
> Oh, the gol-blamed thing will hang in the air
> And energy will never equal MC-square.[35]

Bob Considine reported on June 23 that the correspondents were assembling "a musical review which should set the theatre back at least 50 years." Considine wrote that Laurence was being considered for the role of King Juda, the chieftain of Bikini. "If any cork can be found on the Appalachian Laurence will appear in blackface," Considine wrote.[36]

In the midst of these silly but minor and perhaps understandable behaviors for overachieving journalists trapped together with little to do, Laurence committed something far worse: a major act of brazen plagiarism. On June 25 he submitted a front-page story that he copied, almost in its entirety and without attribution, from a lengthy Operations Crossroads press release describing the Able test. Laurence's story described the arrangement of navy ships dispersed across the lagoon, the schedule of events leading up to the explosion, and the military's plan for reentering the lagoon once it had been declared safe of radioactivity. The article contained no attributions or direct quotes and was replete with unexplained military terms such as "sector axis." Laurence wrote the first nine paragraphs of the story himself and then simply attached pages from the release. Laurence did not even bother to retype the press release. The paper duly ran his article on page 1.[37]

This behavior is difficult to understand and impossible to excuse. Perhaps Laurence felt it was beneath his station to rewrite an official press release that laid out the logistical details of the Able test, but his newspaper was spending a considerable sum to dispatch him to the South Pacific precisely to publish his unique interpretation of events. And Laurence certainly wasn't overworked. Most likely, he was enjoying his time and status as a talking head for other journalists on the press ship and didn't want to tear himself away to write the story.

The *Appalachian* arrived at Kwajalein on June 28 and continued on to Bikini the next day. Once there, the journalists received updated briefings, toured sites in Bikini, and visited an ad-hoc officer's club. Frank H. Bartholomew, a vice president of the United Press news service, later recalled that Laurence issued a dire forecast about it: "The club will be flattened, and those coconut trees opposite the veranda will look as though God's lawnmower had run over them. So, if you fellows value this beer, you'd better bury it." They did.[38]

Laurence committed another act of plagiarism on June 29, writing a lengthy article for the *Times* Sunday review section that sang the praises of military weather forecasters, who had to determine whether the Able test would be safe to proceed. The story started, "Everything is in readiness here for the testing of the effect of an atomic bomb, dropped from an airplane at an altitude of about six miles and made to explode several hundred feet in the air, on a large array of naval vessels." That much, contained on the first page of his dispatch, appears to have been Laurence's own writing. But the second and third pages were copied from a Crossroads press release discussing how test planners determined whether the weather would be favorable. These pages, bearing a few minor editing marks by Laurence, remain in the National Archives repository of journalists' dispatches from Crossroads.[39] Editors at the *Times* cut two-thirds of the story and tucked the rest onto page 3 as a sidebar to a front-page story by the Associated Press.[40] *Times* editors apparently never detected Laurence's malfeasance. This is not surprising; navy communications officials on the *Appalachian* would radio a journalist's story to shore, where it was relayed by Western Union telegram to the journalist's newspaper.

Thus, *Times* editors never saw the original typescript of Laurence's stories, in which the cut-and-paste plagiarism was obvious.

However, some of Laurence's shipboard colleagues did realize what was going on. Edward F. Jones, a correspondent for *Time*, informed his editors that Laurence "makes common practice of filing full text of handouts readily available to [the] Times bureau in Washington." Jones wrote that *Times* executives should be "burning over his cable tolls" from the lengthy plagiarized stories. Jones's comments were intended to be part of the newsweekly's press section, but *Time* never published them.[41]

Laurence was the victim of plagiarism as well as the perpetrator. Another reporter wrote a ten-page account of the Able blast a day before it occurred, drawing heavily and without attribution from Laurence's reporting about the Trinity and Nagasaki explosions. The article described the blast as "brighter, more dazzling, than 100 suns."[42]

With his plagiarized advance stories out of the way, Laurence started planning for his coverage of the Able explosion, in light of his concerns that the *Appalachian*'s transmission systems would be unable to handle the press corps' demands. The Able test was scheduled for 9:30 a.m. Bikini time, which corresponded to 5:30 p.m. the previous evening in New York. Laurence deemed it "very doubtful" that his story on the Able test would arrive in New York in time for the paper's deadline, so he suggested that he broadcast a radio program shortly after the test, which could be monitored in San Francisco and its content telegraphed or telephoned into the New York office for use in the paper.[43] Times officials set up such an arrangement with NBC, which was very excited to have the famous reporter on its airwaves.[44]

On the morning of July 1, the reporters gathered on the *Appalachian*'s starboard side to witness the Able blast, toting protective goggles they had been issued. As Robert Littell of *Reader's Digest* described the scene, "In the chairs, alertly relaxed, with goggles at the ready round their necks, were the working, competitive, deadline press, in costumes I shall always associate with atomic energy—shorts, slippers, sandals, boudoir slacks, pajama tops, galluses; Aloha shorts shouting with hibiscus and jacaranda against backgrounds of magenta and baby blue; bodies colored the seven ages of sunburn; heads under baseball caps, golfing hats; figures in tired khaki shirts, undershorts, no shirts; hairy torsos slung with binoculars; a St. Christopher medal on a chain dangling beneath a two-day beard."[45]

Vincent Tubbs, the lone African American journalist at Operation Crossroads, predicted to Laurence that the blast would sink no ships. "He 'phoofed' me and I crawled into the boat with those expecting the world to be shaken," Tubbs wrote.[46]

At 9:00 a.m., a B-29 Superfortress named *Dave's Dream* dropped the Able weapon over the lagoon. The weapon detonated 520 feet in the air with an explosive energy equivalent of 21,000 tons of TNT, missing the USS *Nevada*, which had been its target. Two ships sank immediately; three others followed within a day. Still others nearby were damaged severely.

Later that day, as previously arranged, Laurence briefly appeared live on a news show on the NBC radio network, along with David Dietz of the Scripps-Howard Newspapers and other naval personnel and witnesses to the Able test. Laurence told his radio audience that the mushroom cloud

Figure 7.1. The first explosion in Operation Crossroads, code-named Able, was from a weapon dropped from the air over Bikini Lagoon. (Library of Congress)

Figure 7.2. Reporters viewed the wreckage of the USS *Independence* two days after the Able test in Operation Crossroads. (National Archives, Record Group 80, 80-G-627512)

seemed smaller than the ones he had seen before. "But I think that is relatively unimportant. The atomic bomb is still the most powerful force on earth. It is still the greatest problem we have facing us." Lawrence E. Davies, the San Francisco bureau chief of the *Times*, hired a stenographer to transcribe Laurence's comments on the broadcast, and Davies transmitted the transcript to New York for publishing in the paper's earliest edition.[47]

Meanwhile, Laurence also wrote a story that he sent to the *Times* through the navy's standard communications channels, for use in later editions of the day's paper. It showed Laurence's trademark dramatic flair, beginning, "The atomic bomb made today its first public appearance on the world stage. It was the fourth atomic bomb to explode in less than a year, but the first three had been strictly private performances shrouded in wartime secrecy. This time the eyes and ears of the world were watching and listening."[48]

The story—which was headlined "Fiery 'Super Volcano' Awes Observer of 3 Atom Tests"—then moved into first-person narrative. "As I watched the pillar of cosmic fire from the skydeck of this ship it was about eighteen miles to the northeast. It was an awesome, spine-chilling spectacle, a boiling, angry, super volcano struggling toward the sky, belching enormous masses of iridescent flames and smoke and giant rings of rainbow, at times giving the appearance of a monster tugging at the earth in an effort to lift it and hurl it into space."[49]

But the *Times* upstaged its star reporter. Although Laurence's article did run on the front page, the preeminent position on the page—the right-hand column—was reserved for a United Press dispatch from Bikini noting that "of the seventy-three 'guinea pig' vessels that were its target only two were sunk, one was capsized and eighteen were damaged."[50] Others noticed the editors' choice. United Press executive Billy Ferguson gloated to his correspondents at Crossroads, "We had Monday morning banner play in [the] New York Times."[51] Still, *Times* officials pronounced themselves pleased with Laurence's work. "Your stuff came through superbly," Catledge told Laurence via telegram. "Used broadcast material in first edition[,] your own special in last." He concluded, "Good job Congratulations."[52] Davies cabled Laurence that he and his wife had "hung on every word."[53]

Through the *Times*'s news service, which distributed its stories to other subscribing newspapers, versions of Laurence's Able story appeared throughout the country, and Laurence was also quoted by other journalists. The *Nashville Banner* cited Laurence as declaring that Able was "the smallest of all the detonations witnessed by him."[54] The *Honolulu Star-Bulletin* paraphrased Laurence as saying that "the explosion we have just witnessed does not compare with previous bursts."[55] Laurence was wrong; Able is now officially estimated as having had a yield of twenty-one kilotons, the same as the Nagasaki explosion.[56]

The judgment that Able was disappointingly small and insufficiently damaging to the ghost fleet dominated most media reporting. "As a show, it was a flop," Littell wrote. " 'Is that all?' I heard people say. 'Nuclear fission!' one correspondent snorted."[57] The journalistic groupthink is perhaps best illustrated by the headline in the *New York Herald Tribune*: "Bomb Fails to Wipe Out Bikini Atoll."[58]

Some blamed the journalistic backlash on Laurence for stoking journalists' expectations about the power of the Able blast. One critic was United Press's Bartholomew, who had buried the officers' club beer because of Laurence's apocalyptic prediction. In an article analyzing the hype surrounding the Able test, Bartholomew did Laurence the professional courtesy of not naming him, but Bartholomew did assign blame to an unnamed "science writer who observed previous atomic detonations" for predicting that the blast "would sweep Bikini bare, that shock and heat wave would rock [the] Appalachian at sea 20 miles away, that [the] glare if not shielded from eyes would have [a] completely blinding effect." (UP editors made the clear reference to Laurence much more vague in the version of the story sent to newspapers, changing it to only "a science writer.")[59]

Clark Lee, a reporter for the Associated Press who was at Crossroads, wrote in his memoir that Laurence had both overpredicted the explosion and exaggerated it in his story. "Laurence heard a 'mighty thunder that seemed to fill all the space,' but to other listeners it seemed a low, distant rumble that scarcely more than whispered against the eardrum."[60]

Laurence explained away the journalists' reaction as ignorant. He wrote in the second edition of *Dawn over Zero*, to which he added material about the Crossroads tests, "To some of the newspaper men aboard, keyed up to the point of expecting the observer ship to be blown out of the water, the spectacle, obscured somewhat by an intervening white cloud, was a disappointment. To me, who could distinguish between the natural cloud and the atomic cloud, the sight was awesome and spine chilling."[61]

Even Robert Oppenheimer poked fun at Laurence. After Able, he told reporters in California that he hadn't expected the palm trees to be leveled because trees at Hiroshima and Nagasaki at the same distance from those blasts hadn't been destroyed. The atomic bomb was "not an anti-palm weapon," the physicist joked.[62] Laurence was a convenient target but probably wasn't deserving of blame. The historian Weisgall has suggested that much of the press disappointment stemmed from the fact that the observers were fifteen or more miles away from Able and that the weather conditions may have muffled the sound and blast.[63]

The editorial page of the *Times* maintained that Able had been a whopper, warning readers not to underestimate the bomb. "One of the few men in the world, if not the only one, who has seen three of the four atomic explosions—William L. Laurence of this newspaper—reported from the press ship Appalachian, off Bikini, that it still inspired in him the terror he first felt in the New Mexican desert a year ago. . . . If there is any inclination to sell short the atom bomb, Don't!"[64]

On July 5 Laurence warned readers not to jump to conclusions. Laurence claimed that the bomb had been dropped farther from the *Nevada* and had detonated closer to the surface of the water than planned. Both errors would have reduced the bomb's destructive force—the first by placing it farther than expected from the closest ship, and the second because a low detonation would have meant that more of its explosive force was absorbed by the ocean rather than by navy ships.[65] However, Laurence again erred, quoting unnamed sources as saying that Able might have detonated as little as fifty feet above the water surface; in fact, the government eventually disclosed that the device detonated at a height of 520 feet.[66]

After Able, Laurence's editors seemed to lose patience with their reporter. Since he left Oakland, the paper had published only four articles by Laurence but numerous wire service stories with details on developments at Crossroads. Editors started trying to manage him from New York. On July 3 (Bikini time) James cabled Laurence with a tip that all the officers on the salvage unit were once enlisted men. "Is that [a] story[?]" James asked.[67] Laurence immediately sent his own cable to military officials seeking to check on the truth of the tip.[68] On July 4 Shephard Stone, the assistant Sunday editor of the *Times*, cabled Laurence from New York asking whether Laurence had any stories to offer for the Sunday magazine. "Have had no reply" from Laurence on earlier inquiries, he chided. "If so please send suggestion or two."[69] Laurence replied with a suggestion for a story about the implications of the results of the Able test for warship design.[70] Stone replied that Baldwin had already covered that topic. "Let['s] forget magazine piece temporaryily [*sic*]," Stone cabled back. "We may have another idea next week."[71] Laurence also wrote James that he was investigating the tip about the salvage vessel but was being forced by the navy to go to Honolulu between the two atomic tests. "Please advise whether to file daily piece in interim."[72] Finally, James loosened the virtual leash, cabling, "Please file only when you[']ve something you think worth while."[73]

Times editors didn't use a story that Laurence filed July 3 about the initial assessment of damage to the ghost fleet by the Able explosion. "From the deck of our ship we can clearly see the *Nevada* standing there as though nothing had happened," Laurence wrote, "her red-painted body and her white and yellow decks marking a sharp contrast to the other ships in the target array." But Laurence then provided a long list of ships that had been heavily damaged or sunk. The *Times* chose instead to run wire copy about the damage.[74]

Sometime after July 6, Jones, the *Time* correspondent at Crossroads, informed his editors that Laurence "has had his shipboard lecturing spiked by his office," using a journalistic slang term that referred to canceling a story. "This week he received [a] wire saying [the] office hadnt [*sic*] heard from him in three days, that he was sent to Crossroads to work not lecture."[75] But Laurence remained unproductive. On July 19, for example, the *Times* had to run brief wire stories about a rehearsal of the Baker test and the arrangement of ships to be used in the test.[76]

During the break between the Able and Baker tests, Laurence and some other journalists spent time at Kwajalein.[77] In mid-July Laurence covered a trip by Blandy to visit Bikini's King Juda to deliver the US government's official thanks for their cooperation with Crossroads. Laurence's story for the *Times* about the trip had a fawning tone: "The visitors came away deeply impressed with the dignity and bearing of these devout, simple folk who not only profess Christianity but practice it in their daily life. Here is an exemplary community where crime is unknown."[78] After visiting the king, Blandy's entourage rode native outriggers to a picnic on an island. For the return trip, Laurence's vessel swamped while making the 300-yard journey. "The first inkling of his peril was a plaintive shout," wrote Kenneth McArdle, a reporter at the *San Francisco Chronicle*. "We looked over and could see only the four-foot mast and small amount of Laurence and the boat structure above water with the native coxswain some distance astern." The coxswain finally pushed the outrigger to shore.[79]

The Baker test, on July 25, used the same type of atomic device as was used in Able, but in other ways the two tests were very different. While the Able bomb was dropped from a plane onto the ghost fleet, the Baker device was stationed underwater on a platform nineteen feet under the lagoon's surface, near the ships. The water prevented the Baker blast from creating a flash of light or loud boom as Able had done. Instead the

Figure 7.3. The second explosion in Crossroads was the Baker test, using a submerged weapon. The dark objects in the water near the explosion are warships left without sailors to gather data on the effects of an atomic blast. (Library of Congress, LC-USZ62-66049)

explosion produced an enormous mushroom cloud of radioactive water and steam.

Much of the press corps defected from Crossroads after Able's disappointment. By the day of Baker, the press corps on the *Appalachian* had dropped from 117 to 55.[80] But the *Times* bucked that trend by adding Baldwin, who arrived at Crossroads on July 21. He was not a science reporter but rather specialized in military strategy and operations, having won a Pulitzer Prize in 1943 for reporting on the Pacific theater of World War II. He and Laurence were personally friendly, but they had sharply different journalistic perspectives and approaches, and their shared coverage of Baker would set the stage for a decade of conflict between the two.

The lengthy articles on Baker by both reporters were each written as if the other reporter were not present. The front page of the *Times* featured a photo of the massive waterspout and mushroom cloud generated by the Baker explosion. The photo was accompanied by a typically breathless article by Laurence that called Baker "the beginning of a new chapter in the story of the atomic age and another landmark in the story of civilization."[81] On the second page of the *Times*, Baldwin began his own story, "The world's fifth atomic bomb—the first ever to explode underwater—racked and strained today the hulls of eighty-seven target ships and boats anchored, moored or beached in Bikini Lagoon."[82] Baldwin and Laurence emphasized different aspects. Laurence examined the blast itself, lapsing into first-person narrative as he described the mushroom cloud: "I watched it reach a distance over five miles, whereas it seems to have climbed to a height of two miles. It never stood still. It was ever changing its shape and color in a matter of seconds so that it was difficult for the human eye to follow all the phenomena."[83] Baldwin, by contrast, focused on how an atomic blast would affect warships. The Baker test, he wrote, simulated an attack on a fleet in a harbor. An atomic bomb exploded in that setting probably would generate more damaging waves and drench crews with more radioactive water than a bomb detonated deep underwater, Baldwin wrote.[84]

In the wake of Baker, Laurence's oracular status continued. The *Honolulu Star-Bulletin* quoted him once again about the blast's impressiveness, "I never saw anything like it. It is the most magnificent thing I've ever seen—a great bomb—greater than Nagasaki or any of the others."[85] (The explosive yield of Baker was approximately twenty-one kilotons, the same as the Nagasaki and Able bombs.) Even Blakeslee was impressed with Laurence's reporting, sending a telegram to his wife asking that she save issues of the *Times* for him.[86]

After Baker, the two *Times* reporters had to pack up to return to New York. The Charlie test, using a bomb detonated deep underwater, was not expected for several months (and in fact never occurred). Before leaving, the two reporters expressed their views on the overall impact of the Crossroads project, and unsurprisingly they reached divergent conclusions. Baldwin got the pulpit first, on August 1. He argued in the *Times* that the two tests would lead to revisions in the design of naval ships, as well as changes in naval tactics and strategy. He predicted that nuclear power could be used in naval vessels within a decade, and he wrote that ships might be more heavily armored so they could better resist an atomic explosion. But the fact that the bulk of the Bikini fleet survived the two explosions indicated that naval power could still function in a world of atomic bombs, he wrote.[87] (DuBridge's prediction in the *Times* had essentially been vindicated.)

Laurence's analysis, published three days after Baldwin's, was far more downbeat. Laurence wrote that the "average citizen" was misinterpreting the fact that the two blasts had created relatively minor damage. That average citizen "had expected one bomb to sink the entire Bikini fleet, kill all the animals aboard, make a hole in the bottom of the ocean and create tidal waves that would be felt for thousands of miles. He had even been told that everyone participating in the test would die," Laurence wrote, ignoring how his own reporting likely had exactly that effect. "Since none of these happened, he is only too eager to conclude that the atomic bomb is, after all, just another weapon."[88]

In truth, Laurence argued, the Bikini experiments showed that amphibious attacks, like those the Allies used in Normandy, would be impossible against a nuclear-equipped enemy, who could use the weapons to repel the invaders. Atomic bombs would be useful against whole cities, or even just to destroy harbors and the ships in them, Laurence wrote. "They must learn to heed the scientists' warning that no defense against the atomic bomb is possible other than the elimination of war and the effective international control of atomic weapons."[89]

The two reporters didn't differ only on the strategic impact of atomic warfare. They also had different takes on radiation. Even before Baker, Baldwin had underscored the importance of radioactivity, and Crossroads confirmed his view. "One of the major lessons of the test, observers agreed, was the necessity of finding some means of protecting ships and their crews against radioactivity," Baldwin wrote. Ships' boilers and cooling systems, which used seawater, could suck radioactive contaminants into the ship along with the seawater, injuring the crew, he noted.[90] And Baldwin the military analyst, not Laurence the science writer, wrote another article reprising the biological and geological findings from Crossroads.[91] By contrast, Laurence's reporting from Bikini largely ignored the issue of radioactivity. As he had with the atomic blasts in Japan, Laurence focused on the destruction packed by the atomic explosion itself—the blast and searing heat—and their secondary effects such as fires triggered by the immense heat—and ignored the long-lasting radioactivity deposited by the explosion.

However, Bikini would demonstrate that radioactivity could not be ignored. Two days after Baker, Bartholomew and a group of other reporters received permission to take a swim on Bikini Beach, where they had buried the beer from the Crossroads Officer Club. "We found the Club still standing, undamaged," Bartholomew recalled. "Once again we peeled off our shirts, this time to dig the beer out of its sandy shelter." After slaking their thirst, they hung a sign on a coconut tree facing the club's building: "Laurence Memorial Grove."[92]

But Able's radioactivity had left an invisible, deadly imprint on Laurence's grove, as it did on the entire island, and continues to do so today despite the natural decay of the radioactive materials left by Able. In 1956, when Laurence returned to Bikini for a test of the hydrogen bomb, he discovered that another sign had been posted at the grove. "Do not eat the fruit of these trees, they are poisonous." Radioactivity from the test had permeated the soil and the trees growing in the grove named for him.[93]

8

Reporter Grade 8

Returning from the South Pacific, Laurence found that the public was hungry for information about the fearsome new atomic technology. People wanted to know how it worked, how the United States should handle the atomic bomb, whether they should worry about their families' safety, what would happen with the Soviet Union. Laurence was more than willing to generate more fame for himself by holding forth.

For example, in December 1947 Laurence appeared on a radio show starring Mary Margaret McBride, a popular radio host. "If we have an atomic armament race it is only bound to end in an explosion sooner or later," Laurence warned. Moreover, he raised the specter of biological weapons, in which one ounce of material could kill "every man, woman and child in North and South America." Developing a biological weapon would be very easy, he averred: "Bacterial warfare can be almost made in a kitchen sink."[1]

Speaking to a teachers' group, Laurence strongly endorsed international control of atomic energy. "There can be no security unless the nations of the world can work out an effective method for the international control of atomic energy, a method that will make certain that no nation, great or small, can manufacture, or possess, atomic bombs, or the fissionable materials in a form that can be easily converted into atomic bombs," Laurence said. "Without such security war becomes inevitable, and if war comes again it will inevitably be an atomic war, a war in which there can be no victor."[2]

The University of Missouri awarded Laurence a medal. "If there ever was a time for great newspapers, for newspapermen, for honest journalism, it is today, because the next few years are going to tell the story as to whether it is all to end up in the great atomic cloud. And it will be up to the newspaper and the journalist of today and tomorrow," Laurence said at the ceremony.[3]

Laurence was seemingly everywhere. In February 1949 he was in Joplin, Missouri, discussing the challenges of detecting radiation from nuclear explosions.[4] Six months later he appeared on "Atomic Report," a television program on Baltimore's WMAR-TV marking the fourth anniversary of the Trinity test. One of the highlights was a lump of fused sand from the Trinity test site, which a Geiger counter still showed as radioactive four years after the blast.[5]

Even his theological views were of interest. In August 1949 Laurence was a guest on the *Eternal Light* radio show, produced by the NBC Radio Network in conjunction with the Jewish Theological Seminary in New York, discussing the Old Testament prophet Amos. Laurence saw Amos as warning that violating natural law would lead to destruction. "Anyone who transgresses the moral order eventually is wiped out because nature takes care of him because it goes against the very act of nature," Laurence said on the broadcast. He saw a parallel to the danger of depending on the might of the atomic bomb. Instead, he said, America should rely on the strength of its democracy. "If we forget that and rely only on physical force then that would lead to the best way of losing democracy."[6]

The *Saturday Evening Post* tried to leverage its relationship with Laurence. The magazine offered a collection of "valentines" to various readers and authors in a full-page ad in its February 14, 1947, issue. Laurence's valentine read: "To William L. Laurence, 'the best scientific writer in America,' for writing such a remarkably prophetic article about atomic energy 'way back on September 7, 1940. Also, to the same Mr. Laurence for writing the first clear and comprehensive account of the future of atomic energy (April 13, 1946, Post)."[7]

At the same time, his journalistic productivity was taking off. Although Laurence had published only a modest forty-six bylined articles in the *Times* in 1946, the next year he shot up to seventy-seven articles. In 1948 he published a hundred bylined articles, and in 1949 he reached 108 articles—the most of any year in his career at the *Times*. At the same time, he was writing less about atomic energy; of his one hundred articles in 1948, for example, only eight dealt with atomic energy.

But Laurence's fame did have limits. Laurence and his wife were arrested by British occupation forces in Germany in July 1947 as the couple traveled by train to cover a microbiology conference in Denmark because they did not have necessary visas needed for their intermediate stops. Dutch officials had ignored the missing visas and had permitted the couple to proceed by train into British-occupied Germany, but British officials there were not so lenient. They forced the Laurences off the train and refused to permit them to use the telephone or telegraph, and the two were expelled to the Netherlands under guard in a freight train. They finally flew from there to Copenhagen.[8]

Laurence maintained a special relationship with Albert Einstein, whose research made an appearance of sorts at Laurence's sixtieth birthday party in 1948. Florence gave her husband a chair from the Harvard University library, and guests included aviator Jimmy Doolittle, the Arctic explorer Peter Froechen, and cartoonist Bill Mauldin. The cake was adorned with the famous equation $E=mc^2$. "The nucleus of the atom was iced in multicolor dots of frosting, centered with seven different color circles, to represent the seven shells around the atom bomb. This was explained as the cake, which was good to eat as well, was being cut," one reporter related. A second cake had a picture of their dachshund and the name "Einstein," in reference to their dog. Ever serious, Laurence warned partygoers that Russia was five years away from developing its own atom bomb.[9] Laurence vastly underestimated the Soviets, as it turned out.

The next year Laurence returned the favor, in a way, by writing a news article noting Einstein's seventieth birthday. "The man revered above all others by scientists the world over as the outstanding intellect of his generation, who was acclaimed as one of the immortals of history at the age of twenty-six, will observe his reaching the Biblical three-score and ten

Figure 8.1. Laurence claimed a friendship with Albert Einstein. Laurence and his wife, Florence, visited Einstein's home in Princeton, New Jersey, in January 1950. (*New York Times*)

quietly at his modest home in Princeton."[10] Laurence's fascination with Einstein left the reporter vulnerable to being influenced by public relations tactics when Einstein published a new edition of his book *The Meaning of Relativity* in 1949. Einstein's publisher, Princeton University Press, hyped the book at the annual conference of the American Association for the Advancement of Science as a major advance in Einstein's efforts to develop a unified field theory. Always willing to write a slavish article about Einstein, Laurence took the bait with a front-page story reporting that Einstein's theory "attempts to interrelate all known physical phenomena into one all-embracing intellectual concept, thus providing one major master key to all the multiple phenomena and forces in which the material universe manifests itself to man."[11] One scholar has noted about Laurence's hyperbole: "Three 'alls' in one sentence describing a theory of everything ought to have alerted even the most credulous that something was amiss."[12] But other journalists covering the meeting also jumped on the bandwagon, including science reporters from the Associated Press and the *Christian Science Monitor*.[13] Despite the fawning coverage, Einstein had not resolved the problem of unifying theories, and in fact he never did so.

Meanwhile, big changes were afoot in the US nuclear weapons program. Without the wartime urgency that had driven the Manhattan Project, the Los Alamos atomic design lab lost many of its talented personnel, and questions swirled about the direction of weapons research and how best to develop new weapons. The United States resumed nuclear testing in April 1948 with Operation Sandstone, a series of three explosions at Enewetak Atoll in the South Pacific designed to test new designs to replace the rudimentary models used on Hiroshima and Nagasaki. For example, the model used on Nagasaki, called Fat Man or officially Mark III, was heavy because of the amount of high explosives needed to set it off. Its batteries had a short life, and the bomb casing's shape reduced the accuracy of delivery.[14] Sandstone's three tests of new design features made the existing weapons in the US arsenal obsolete.

Contrary to his reputation as a leading journalist in atomic issues, Laurence wrote nothing about these issues. In one article in December 1947, Laurence briefly mentioned construction for nuclear testing at Enewetak Atoll, but he reported nothing else about Sandstone's plans or the changes rippling through the personnel and technology of US nuclear weaponry; the stories that the *Times* ran about Sandstone were all from the Associated Press. Unlike Crossroads, Sandstone was not open to reporters; as media scholar Beverly Deepe Keever has noted, the *Times* made no editorial objection to this policy. On May 18 officials held a briefing in Honolulu. The *Times*, in a brief unbylined article, reported that there had been no air drop in the test but did not characterize the Sandstone results. Fortunately, other journalists were on the story: The Associated Press reported that the Sandstone tests demonstrated how the atomic weapons worked, that they were of an improved design, and that they were successful. The International News Service also reported that Sandstone used weapons of a new design. By contrast, the *Times* would not tell its readers about the improvement in design for two months, after the AEC issued an official report—and that story was not written by Laurence but rather by Anthony Leviero.[15]

The hottest question of the moment was whether the Soviet Union was an atomic threat to the United States, or when it might pose one in the future. Laurence was placing less faith in the Soviet Union's nuclear capabilities than many others. While speaking to a training class for science teachers in 1948, Laurence predicted that the Soviet Union would not produce its first atomic bomb until 1952 (a poor prediction, as the first Soviet nuclear test occurred in 1949) and would need at least twenty-five years to match the size and quality of the US stockpile. At the same event, Baldwin—the military reporter at the *Times*—suggested that less time would be required. The *Times* tried to paper over the differences of opinion between Laurence and Baldwin with an unbylined story on speeches by both, asserting that they had agreed "that the United States would have a protracted breathing space before Russia could match it in quality and quantity of atomic bombs."[16] In retrospect, the other half of Laurence's prediction was more accurate: the two stockpiles did not have equal numbers until 1978, according to Western estimates.[17]

Officials at the Atomic Energy Commission were unhappy with what they saw as Laurence's lax use of classified information he had obtained during his service for the government on the Manhattan Project. In July 1948 Laurence asked Morse Salisbury, head of the AEC's public information department, to comment on the draft of an article. Laurence had used various facts and assumptions about the US program's pace of manufacturing nuclear weapons to conclude that the United States had no more than three hundred nuclear weapons in its stockpile and probably no more than two hundred. (Today, scholars estimate that the United States had fifty nuclear weapons at the time.)[18] Salisbury refused to comment on Laurence's article, writing that he didn't know the rate at which weapons were being manufactured; even if he did, it "would be prejudicial to security" for him to comment. But he also warned that Laurence might be veering close to classified information he had learned during his Manhattan Project days: "Your publication of such speculation is a matter for your own decision. Involved, I presume, is the fact that your connection with the Manhattan District may have resulted in your having data bearing on the calculations which you undertook in preparing the draft."[19]

Laurence proceeded with the article anyway, and it was published on November 6, 1948—in the *Saturday Evening Post*, not the *Times*. He told the magazine's readers that the United States had significant breathing space in the new A-bomb arms race with the Soviet Union because of that nation's limited industrial capacity: "We still have at least four years, as of today, before Russia can begin producing A-bombs of her own."[20] That is, Laurence was predicting no Soviet fission bomb capability before late 1952. At the time, the Central Intelligence Agency was estimating that the Soviet Union could not complete its first atomic test before mid-1950. Both Laurence and the CIA were wrong. The Soviet Union exploded its first atomic bomb—a plutonium device that had been copied directly from the US's Fat Man—on August 29, 1949.[21]

However, the Soviet Union did not announce that achievement to the world. Rather, on September 23, 1949, President Truman revealed that the United States had evidence that the Soviet Union had conducted its first nuclear explosion. Unlike other journalistic colleagues at the *Times* and the *Washington Post*, Laurence in his front-page article stressed not the fact of the explosion but its timing. "Though the scientists have predicted its coming, it came at least three years sooner than was expected," Laurence wrote. He erroneously predicted that the Soviet Union could have a stockpile of fifty atomic weapons within a year.[22]

This view was anathema to the Truman administration, which was pursuing a balanced federal budget and thus could ill afford a costly atomic weapons arms race with the Soviet Union. Administration officials were predicting a slow growth in the Soviet atomic stockpile. In the weeks after the announcement of the first Soviet A-bomb explosion, other US media organizations such as *Life* magazine followed Laurence into a drumbeat for heightened American atomic military strength. Soviet leaders also began to suggest they would be able to reduce or eliminate the US advantage in nuclear stockpiles.[23]

As Laurence continued to build his reputation, he sorely wanted to be able to boast of holding a degree from Harvard University; the fact that he didn't vexed him terribly. After returning from World War I, he instead had earned a bachelor of laws degree from Boston University in 1925. While considering a career as a lawyer, he did something that permanently broke his already fraught relationship with Harvard: in the fall of 1925, Laurence was caught trying to take an examination in elementary German for a Harvard College student whom Laurence had been tutoring.

The proctor knew the student Laurence was impersonating, so Laurence was caught.

Harvard never forgot the infraction, even though Laurence asked time and again for the degree that he believed he had earned in 1917. In 1937 Laurence had worked through a classmate to recruit a Harvard professor to ask A. C. Hanford, then the dean of Harvard College, about Laurence's situation. Hanford deflected the inquiry, assuring the professor that if he reviewed Laurence's record "you will agree that Harvard College is much better off without having Mr. Laurence listed among her graduates."[24] In 1947 Laurence enlisted Ralph Lowell, a fellow member of the class of 1912 who had risen to become a member of Harvard's Board of Overseers (one of its two governing bodies), to intercede with the university administration on his behalf. After a speech in Boston, Laurence met at the Harvard Club with a group of Harvard alumni including Lowell, who was thoroughly wowed. "Bill is one of the plainest looking men in the world, but when he starts talking about the atom bomb and its menace to the future of the world he becomes almost religious or ecstatic as he speaks plainly, looking over the heads of his audience without any notes," Lowell wrote in his diary.[25] Lowell took up the matter of Laurence's missing degree with Wilbur J. Bender, a senior administrator at Harvard. But the Harvard administration stuck to its guns. Bender, who later would become dean of admissions at Harvard, replied that Laurence had had financial problems as a student but also alluded to the cheating incident without describing it explicitly. Bender wrote that Laurence "was involved in a most unsavory incident which, I believe, should end all chance of his getting a Harvard degree."[26]

A few months later, Laurence himself wrote to Hanford on *New York Times* letterhead to request a transcript of his academic record "to Commencement, 1915" (an endpoint that conveniently would have excluded the cheating incident). He pointed out that he had been written up in *Who's Who in America* and *Current Biography*. "As I am now approaching three-score years, I naturally am anxious to put my house in order."[27] Laurence's allies sent a copy of Laurence's letter to Harvard's president, James B. Conant. The situation could have been personally awkward for Conant, because he knew Laurence both from his days as a student and from his Manhattan Project work. (Laurence had quoted Conant at least eleven times in his *Times* reporting.) Members of the Harvard Club in

Manhattan had rejected Laurence's application for membership because he did not have a degree from Harvard; Laurence and his friends hoped that a copy of the letter, with a copy of his Harvard transcript, could win him a membership.[28] Bender, who had succeeded Hanford as dean of Harvard College, pulled no punches. He noted Laurence's troubled history with creditors and added: "Much more serious was the incident in which you were involved in 1925. . . . Apparently you wrote the examination, wrote the student's name on it, tried to pass it in as the student's work, and were detected."[29] Laurence wrote back, again on *Times* letterhead, to argue that his career had proven that he deserved a Harvard degree. Laurence argued that "any transgressions" a person has committed should "be placed in the proper perspective against the background of the rest of his acts throughout his lifetime."[30]

Bender sought to reassure Laurence while yielding no ground on the issue of the degree. "We have no intention to broadcast this information or to do anything to cast a shadow on the splendid record of useful achievement you have made in the years since you left Cambridge. I should think that you can take deep satisfaction in that and let the Harvard record rest in peace."[31]

But as any dean who has worked with aggrieved students might have anticipated, Laurence did not drop the matter. A month later, Laurence wrote again to Bender raising the possibility that Bender's disavowal of punishment "can be interpreted to mean that Harvard may at last grant me my degree, the further withholding of which I cannot help but regard as continued disciplinary action."[32] He also went above Bender's head by writing to the provost, who asked Bender about the situation but apparently was satisfied with Bender's explanation of how it was being handled.[33]

Bender tried again to quash Laurence's hopes. "It is the policy of the College not to reopen cases like yours which would be painful for all concerned and where it is no longer possible to get at all the facts, since those in the College who know most about the circumstances are no longer able to express their views."[34] But Bender noted that he was new in the dean's position and would have to inquire further for a final answer. Six weeks later, Bender reported to Laurence that the administrative board had examined his case thoroughly and voted not to recommend a degree for him, "thus closing the issue finally." Making an exception for Laurence would be inappropriate, Bender wrote, "because your action in impersonating a

Harvard student in an examination would have led to your expulsion if you had still been in College."[35]

Laurence responded with a three-page, single-spaced letter that finally offered his explanation of the cheating incident more than two decades earlier. "I had tutored the undergraduate in question all that Summer, and at great sacrifice to myself and without fee, because he had convinced me that his entire future depended on it and that I was the only one who could help him." The student asked Laurence to drive him to the exam to give him moral support. As they sat in Laurence's car outside the lecture hall where the exam was to be administered, "in a sudden emotional outburst, he informed me that he was sure that he could not pass the examination and that unless I substituted for him he would commit suicide right on the spot with a gun he had in his possession." Laurence said he "chose the lesser of two evils." Despite the cheating incident, Laurence maintained that Harvard owed him a diploma, because he had fulfilled all his degree requirements and paid all his debts to the college in September 1915.[36]

"This completes my final soliloquy," Laurence concluded his letter, which indeed is the final letter from him in his academic file. "And as the curtain slowly descends, I make a profound bow to Harvard and wish her well. May moss-covered error never again moor her at its side, as it has done in this case. And as a final curtain speech may I add that I always was and shall ever remain a loyal son of Harvard, true to her traditions and proud to be regarded by the outside world as one of her products of which she need not be ashamed."[37]

In later years Laurence wasn't always fully forthcoming about his academic pedigree. A 1955 article about Laurence in the *Times* employee newsletter, *Times Talk*, stated he graduated from Harvard in 1915 and said nothing about Boston University. Laurence presumably was the source of that information.[38] In 1970 he told an interviewer, "After four years of philosophy in which I graduated with honors, I found that getting a job teaching philosophy wasn't easy, especially for a Jew."[39]

In a 1964 oral history, Laurence ascribed his lack of a Harvard degree to his unpaid bills at the university and a dean's vendetta against him, making no mention of the cheating incident that occurred long after the dean's departure and strongly influenced the university's decision to withhold his degree.[40] Meanwhile, Laurence did not publicly emphasize his degree from Boston University, although that university has taken some pride in counting him among its alumni. In a 1949 issue of its alumni magazine celebrating alumni who had connections to atomic energy projects, Laurence featured prominently. The magazine called him "the most outstanding science reporter of our time."[41]

Laurence's relationship with the *Times* was little better than his relationship with Harvard. By February 1949 Laurence believed that he had demonstrated an ability to find and report important scientific stories that other journalists would miss, but he felt that his arrangement at the *Times*—where he reported to the city desk—was holding him back, so he wrote Edwin James to propose a major change. "The developments in the field of science have become of transcendent importance in the lives of all of us," Laurence wrote. "During the nearly twenty years in which has been my privilege to work on the Times I have raised the level of science reporting to heights never attained before, having set standards in the field that have won universal admiration and respect." He claimed that he had "won more prizes than any other individual newspaperman."[42]

Nevertheless, "I am still subject to routine and trivial local assignments, which are not only undermining my morale and self-respect but also prevent me from utilizing my knowledge, experience and skill to the full, in giving the Times the kind of coverage of important science developments, in this country and abroad, for which I, above all others, am uniquely qualified."[43]

Despite his accomplishments, he wrote, "I find to my profound regret and disappointment that I am still largely where I was when I first came to the Times in the Spring of 1930." On salary, this was not true. Times records show that he was hired in 1930 at $100 per week; when he wrote his 1949 memo, he was earning $231 per week. Nevertheless, Laurence complained, "It is rather ironical that the highest salary rises that have come to me have not been merit rises but general ones, given to everyone," he wrote. "My salary is still rather low compared with my high standing in the profession."[44]

Laurence also proposed to James that the *Times* create a dedicated science news department, separate from the city desk. That would allow him to cover all scientific fields, report on scientific conferences, and visit labs and research facilities. "In this manner I would supply the Times with a constant stream of exclusive stories such as only I could get." He also would look toward his retirement. "I would make it my business to be on the lookout for promising young men whom I could train in the art (and it is a high art) of finding and reporting news in all major fields of science accurately and interestingly, with an eye also for the social and political implications of future scientific developments, so that they could step into my shoes when I am no longer around."[45]

After the *Times* failed to do as he asked, Laurence turned to the Atomic Energy Commission, asking in April 1949 for a so-called Q clearance, which permitted access to classified information about atomic weapons, usually granted to someone who was working on a weapons project for the government or a contractor. First, he was proposed for the clearance by Michael Amrine, a writer specializing in atomic power coverage (probably at Laurence's instigation), but the AEC's Corbin Allardice opposed the idea because it would give Laurence an edge over other journalists. Next Laurence himself proposed the idea to AEC's Morse Salisbury, asking for the clearance so he could research a series for the *Times* on developments in atomic energy since 1945. Salisbury pointed out that the clearance would bar him from disclosing any classified information he learned in his reporting. Laurence countered with a proposal that he could conduct the research for the AEC rather than the *Times*. The clearance never was granted.[46]

Back at the *Times*, Laurence's complaints apparently were effective enough to draw him a twenty-five-dollar weekly (11 percent) merit raise in May 1949.[47] But he was still working for the city desk, and Laurence didn't stop complaining about that. In June Laurence again raised the issue of working independently from the city desk. James advised Sulzberger against the idea.[48]

In July he appealed directly to Sulzberger to have his job reclassified from "Reporter Grade 8" reporting to the city desk to "Science Reporter at Large" independent of the desk, as were some other senior reporters such as Baldwin. Laurence pointed to an article he had published just that day about reluctance among some American officials to share information on atomic energy programs with Britain and Canada, despite the fact that researchers in those two nations already were well acquainted with the technology through their work in the Manhattan Project. Laurence wrote that the article "speaks for itself" as evidence of the reporting that he would produce for the *Times* if he were reclassified as science reporter at large. The article was based on his "intimate first hand knowledge." Given his status as a Reporter Grade 8, he probably would have been occupied with "some routine assignment that any other Grade 8 reporter could have done."[49]

In his memo Laurence cited his landmark May 5, 1940, story on fission as another example of how the city desk was slowing him down. "I can assure you that had my status been different at the time I could have written that story at least two months earlier than I did, and I thanked my lucky stars at the time that it remained exclusive as long as it did."[50]

Laurence's memo unleashed consternation among his supervisors when Sulzberger asked for their comment. James retorted that Laurence had been able to write the story on the British precisely because he was not on a "picayune assignment." And James reported to Sulzberger that Laurence recently had requested to write articles about atomic energy that would be released to all the press, as his Manhattan Project releases had been. James rejected that idea but suggested that Laurence should see if the AEC would be willing to pay for them. "I have heard nothing from Mr. Laurence since then about the project."[51] Presumably James did not know about Laurence's spurned offer to write a history for the AEC.

In response to a draft response by Sulzberger, James wrote a second, longer memorandum that suggested that Laurence had plenty of freedom. James calculated that in the preceding 151 work days, Laurence had been out of town on "assignments of his own selection" for 45 days and in the office on assignment for 35 days. Laurence had had no assignments at all on 30 days. James further suggested that Laurence could be a bit quicker out of the gate. The information on which Laurence had based his July 20 story about the British had been available a week earlier, on Friday, July 13, James reported. On that day, Laurence had no assignment. On the following Monday, he had no assignment. Only on Tuesday, July 19, did Laurence voice the idea for the story and receive permission to proceed on it. James still opposed the idea of giving Laurence greater freedom. "I think he is trying to put over something which would result in a weakening of the organization of the News Department," James wrote.[52]

Sulzberger then responded to Laurence, pointing out a recent compromise on salary and office space. He reminded Laurence that *Times* management had been firm that he should report to the city desk and that "we did not wish to take on several extra persons in order to create a so-called Science Department."[53] For an individual with much larger ambitions than just working at the *New York Times*, Sulzberger's response was an invitation for Laurence to continue to spread his wings. He would do exactly that, to the *Times*'s regret.

9

The Elixir of Life

His joints crippled by an advanced case of rheumatoid arthritis, the middle-aged man in a movie prepared by Mayo Clinic researchers slowly and painfully hobbled down a set of steps. But only eight days later, after receiving a new medicine developed at Mayo, he briskly descended the steps for the movie camera. He even sprinted a short distance. Five days after the man stopped receiving new doses of the wonder drug, first called Compound E and later rechristened cortisone, the movie showed that the man had relapsed and was barely able to walk.

In a standing-room-only auditorium in Rochester, Minnesota, on April 20, 1949, Laurence furiously took notes on the riveting movie, which depicted the amazing experiences of more than a dozen research subjects who had received cortisone injections at Mayo. Laurence also listened carefully to detailed presentations by two Mayo researchers. As the only journalist present, having essentially crashed the party, Laurence then rushed to a telephone to dictate a front-page story that announced a new era in treating the debilitating disease of rheumatoid arthritis. In one stroke, he reminded his journalistic colleagues around the world that he was not just adept at covering atomic energy; he was one of the preeminent journalists covering all of science.

Cortisone is a steroid produced in the adrenal glands, small masses of tissue that sit atop the kidneys. In the 1930s the Mayo Clinic's Edward Kendall had identified six different chemicals produced by the adrenal glands, which he called Compounds A through F. Through the 1940s he and Mayo's Philip Hench began to zero in on Compound E as a possible treatment for rheumatoid arthritis, but the chemical was available in such minute quantities that its effectiveness was impossible to test. Compound E was costly and time consuming to manufacture from scratch, but the pharmaceutical company Merck & Co. began to try to do so in the 1940s. By March 1949 the company had produced enough to enable the Mayo researchers to test Compound E in five communities around the country, in collaboration with eminent rheumatologists who were sworn to deep secrecy about the nature of the project.

The positive results were dramatic. Knowing that the tests would trigger intense public interest, officials from Mayo and Merck worked out a detailed plan for releasing information about the research, centered around a closed-door presentation by four Mayo researchers on April 20, 1949. Mayo and Merck would jointly announce the paper three days later. The researchers also agreed to a gag order: "In so far as possible, no interviews concerning compound E will be granted by any author," the plan stated.[1]

But their carefully crafted media plan didn't take account of Laurence, who prided himself on ferreting out obscure but important scientific news stories. On April 19 Dr. Clarence Kemper, a Mayo rheumatologist visiting New York, bumped into Dr. Howard Rusk, a physician and expert in arthritis who also was an associate editor at the *New York Times*.[2] "When is Mayo's new arthritis discovery going to be announced?" Rusk inquired. Kemper temporized, probing Rusk in order to ascertain whether he had been granted access to the highly confidential project. Rusk recited enough details to mislead Kemper into thinking that he was indeed in the know, so Kemper replied, "As a matter of fact, it's going to be presented tomorrow night."[3]

Rusk raced to a telephone to tell Laurence but couldn't reach him, so Rusk told the *Times* city desk: "Get in touch with him, tell him to take the next plane to Rochester, Minn., and call me when he gets there."

In Detroit to cover a scientific conference, Laurence thought that he had plenty of time to get to Rochester, so he made a dinner date with Alton Blakeslee, a science reporter for the Associated Press and the son of Laurence's longtime colleague, friend, and competitor Howard W. Blakeslee. But then Laurence learned that he had to be at the airport within fifteen minutes to get to Rochester that evening. He didn't check out of the Detroit hotel or pack any of his belongings before dashing to the airport; he didn't even inform Blakeslee that he was breaking their dinner date. Eventually, Blakeslee became so concerned about Laurence's failure to show up for dinner that he called the Detroit police to report him as missing.[4]

Laurence did make it to Rochester that night. The next morning, to the surprise of Mayo officials, the reporter presented himself at the clinic and asked for an advance copy of the manuscript to be delivered that evening and for permission to cover the presentation itself. That caused two problems for the Mayo officials. One was the issue of Laurence's desire to be present at the presentation. Mayo's media plan had envisioned the April 20 meeting as one for researchers alone, and no journalists had been invited. But given Laurence's highly praised coverage of the atom bomb project, Mayo officials judged that "his reliability was above reproach." They decided to allow him to attend the staff meeting while not informing any other journalists about the event.[5]

The second issue stemmed from Laurence's request to read the paper before it was delivered. Science and medical journalists customarily received advance copies of papers so they could digest the technical material. Officials at Mayo, however, held to the principle that scientific papers should be communicated first to scientists, so they refused. "The denial was in keeping with policy, and if there is anything the Mayo Clinic does well, it's 'adhere to policy,'" one observer has written.[6] Laurence raised a stink about the decision. Two Mayo researchers recalled his reaction, almost a half century later: "Mr. Laurence considered this an affront to his integrity or a reflection of Mayo naivete or both and he never forgot it as long as he lived."[7]

The cortisone presentation on the evening of April 20 was a sensation. "Every member of the staff of the Mayo Clinic and Mayo Foundation who could be there was present," Kendall recalled in his memoir. "Every seat in Plummer Hall was taken, and chairs were placed in the aisles, but many sat on the window sills and even around the platform. Others stood along the sides of the hall and even out to the elevators."[8] Sitting in the auditorium's front row, Laurence provided an opening for a light touch from the master of ceremonies. "He raised his arms to the physicians packing the room, then nodded politely to Laurence. 'I'm happy to see that so many of you are interested in uterine tumors,' he said."[9]

The motion picture documenting the transformative benefits of cortisone mesmerized Laurence. "One after another we saw tortured men and women, young and old, some bedridden, some in wheel chairs, suffering excruciating pain at the slightest touch or more," he later recalled. "One by one we saw these same men and women—sixteen in all—transformed into smiling, happy human beings, walking jauntily, performing exercises, acting in every way like normal people."[10] Laurence wasn't the only person moved by the drug's transformative effect. Another person present recalled, "It was like God had touched them."[11]

After the movie, Hench delivered a twenty-minute presentation about Compound E, followed by a fifteen-minute talk by Kendall. "When the lights went on the audience of scientists gave Doctors Kendall and Hench an ovation such as is seldom heard. All of us knew that we had been privileged to witness one of the important landmarks in man's eternal battle against disease and suffering," Laurence later wrote.[12]

As soon as the presentations were over, Laurence quickly composed his story and dictated it over the phone to the *Times* newsroom in New York, after the early editions had been printed but in time for the late edition. "Preliminary tests during the last seven months at the Mayo Clinic with a hormone from the skin of the adrenal glands has opened up an entirely new approach to the treatment of rheumatoid arthritis, the most painful form of arthritis, that cripples millions," Laurence's story began.[13] Laurence highlighted the visual evidence provided by the Mayo movie but also quoted Hench as warning against "too much optimism" because of the early nature of the study. The front-page placement of the story and Laurence's strong lead signaled that Compound E was an important scientific step, but the rest of Laurence's story appropriately indicated that using Compound E as a treatment for the millions of arthritis sufferers was some time in the future, if indeed it would happen at all. Although they had been nervous about admitting Laurence to the presentation, Mayo officials judged that "Mr. Laurence wrote an excellent report without having had any advance information."[14]

Other newspapers ran only wire service stories about the research. The *Washington Post* and the *Chicago Daily Tribune*, for example, first carried only an AP report that rewrote Laurence's article.[15] The United Press distributed a story that indicated that Compound E was not a cure for arthritis but might be a "highly effective" treatment for the disease's symptoms.[16]

The Mayo Clinic's local afternoon newspaper, the *Rochester Post-Bulletin,* was caught flat-footed by Laurence's scoop. On the afternoon of the day it appeared, the Rochester paper ran two front-page articles on the Compound E finding. One article focused on Compound E, stating that the steroid had "significant promise in the treatment of rheumatoid arthritis." The second article provided brief biographical sketches of the four local researchers who had participated in the project. Neither article bore a byline, neither had any direct quotes, and neither attributed information to any named sources. Neither article mentioned Laurence's reporting.[17]

Laurence was still in town and still on the story. The day after the Mayo meeting, he filed another dispatch, this time relying on the text of the scientific paper. Laurence included details such as the ages and genders of the subjects and the duration of treatment. He also fanned the excitement about the results, stating that "all present" at the Mayo meeting "agreed that they had watched a modern miracle and had been privileged to attend one of the great turning points in the battle of man against disease." The paper's editorial page agreed, calling it an "epoch-making discovery—another great victory in man's battle to conquer disease." Two days later, Laurence's colleague Waldemar Kaempffert reprised the news about Compound E in a generally favorable article but predicted that researchers would need at least ten years to determine whether the drug would be effective in treating arthritis.[18]

Other science journalists were slower out of the gate. It was more than two weeks until the *Washington Post*'s science writer, Nate Haseltine, offered his own take on the research. Haseltine noted that Compound E was remarkable but also stressed the limited quantities of the drug and the fact that daily injections would be required to maintain its effect in a patient.[19]

Journalists focused on the upcoming seventh International Congress on Rheumatic Diseases, in late May in New York, at which Hench was expected to discuss the drug. On May 29 the *Times* ran two pieces previewing the conference; Rusk predicted that "undoubtedly the most dramatic report of the Congress" would be the one from Hench and Kendall about Compound E.[20] The *Washington Post*'s Haseltine also wrote a curtain-raiser for the conference, saying that Hench was expected to provide "the first full report" on clinical tests of Compound E.[21]

Once again, the Mayo movie of arthritics treated with Compound E turned heads at the conference. One reporter described it: "Men and women who had been bent and stiff, barely able to walk without help, were shown stepping friskily up a flight of stairs a few days after their first injection."[22] The Associated Press reported that Compound E "has worked on every case of rheumatoid arthritis."[23]

Laurence continued to outpace the competition with a lengthy article from the congress noting that Compound E also had been used to treat another form of arthritis, called Strümpell-Marie spondylitis (today known more commonly as ankylosing spondylitis). Laurence also reported that the US military encouraged development of techniques to produce cortisone during World War II in reaction to rumors that Germany was using the hormone to enhance the performance of its military pilots. A day later, Laurence reported on another paper delivered at the conference by researchers at Harvard Medical School suggesting that cortisone also might be effective in treating epilepsy. After the conference, Kaempffert filed a review of the conference agreeing that cortisone did "seem to perform what can be described only by that much-abused word, 'miracle.'"[24]

At this point the issue of finding adequate sources of cortisone started to fascinate Laurence, both as a news story to be covered for the *Times* and as a problem to solve for the public interest. He discussed the matter with Louis F. Fieser, a hormone expert at Harvard University, at a meeting at the National Academy of Sciences in Washington, DC. Fieser told Laurence about a plant whose seed contained a substance that was similar to cortisone, but he couldn't remember the details, so Laurence searched for them himself. Chemists at Merck told him that the substance was called sarmentogenin. Mayo's Kendall referred him to Walter A. Jacobs, a researcher at the Rockefeller Institute in New York, who said that in 1915 Rockefeller researchers had found a substance called sarmentogenin in some seeds. Laurence concluded that this could be the key for widening access to cortisone, because the series of thirty-five chemical steps required to synthesize the complex cortisone molecule from scratch in large quantities was deemed impractical, while sarmentogenin could be converted into cortisone with only eighteen steps—a huge savings in time and cost.

A major question remained: which vine produced the seeds with sarmentogenin? The seeds analyzed in 1915 had been labeled as from the species *Strophanthus hispidus*, but scientists later concluded that the label was incorrect. A decade later, the Rockefeller researchers found sarmentogenin in a batch of seeds labeled as *Strophanthus sarmentosus*, a related but different species. They didn't know where it had been grown, but they did tell Laurence that German scientists had studied sarmentogenin. Fluent in German, Laurence scoured German-language scientific journals at the New York Public Library and obtained pictures of *S. sarmentosus* plants from the New York Botanical Garden. Eventually, he learned that Swiss scientists were pursuing the vine. He unearthed a German-language report on their work, and a source in a pharmaceutical company told him that the Swiss botanists had been trying to gather *S. sarmentosus* seeds in Africa.[25] One author wrote: "Laurence seems to have been the only person in America at that time to associate the possibilities in the *Strophanthus* plant with the pressing need for cortisone."[26]

By June 1949 Laurence was ready to write the story, but he realized that the vine grew only in equatorial Africa, and he worried that publicizing the value of the vine might encourage Britain, which controlled the region, to restrict exports to create a monopoly on the plant. Recalling that the United States had struggled under foreign embargoes of rubber and quinine during World War II, Laurence worried that his reporting could restrict America's access to cortisone.

Florence suggested that Laurence inform President Truman about the situation. When her husband hesitated, she telephoned the White House press secretary, Charles Ross, who expressed interest in the issue. Encouraged, William Laurence wrote Ross a four-page, single-spaced letter on June 21, arguing that cortisone injections were vital for "7,000,000 arthritic cripples in the United States." He argued that cortisone also would be useful in treating and even preventing rheumatic fever and mental illness. As a result, he wrote, "many millions of Americans, of all ages, will need it." Laurence wanted to alert Truman to the situation so that he could take steps "to protect the interests of our citizens, so that no foreign nation will have a stranglehold on a matter so vitally important to the health of our people." He requested a personal meeting with the president. "I am convinced that the assurance of an adequate supply of the seed of the rare African plant is just as important to our public health as the development of the atomic bomb was to our national security."[27]

Ross was sold. The next day, he telephoned Laurence to schedule a meeting with Truman at 11:15 a.m. on Tuesday, June 28. Laurence told no one at the paper; he simply asked for June 28 off. Meanwhile, Laurence prepared a five-page confidential memorandum for Truman.[28] The document opens with a brief description of the *Strophanthus* plant and the fact that its seeds are the only naturally occurring source of sarmentogenin. Mayo's discovery of cortisone earlier in the year had transformed sarmentogenin into "one of the most valuable plant products in existence" because sarmentogenin was the only available precursor for synthesizing cortisone, Laurence wrote.[29]

Cortisone could be manufactured from the bile of cattle, but Laurence estimated that 14,600 heads of cattle would have to be slaughtered to supply cortisone for one patient for a year. By contrast, one ton of *Strophanthus* seeds, grown on one acre of land, could be used to manufacture the same amount. "This opens the possibility of a very profitable vast new industry in the undeveloped areas of the world, such as the equatorial regions of Africa, Brazil, Venezuela, the Philippines and the East Indies." He argued that a crash project cultivating the plant would align with Truman's own Point Four Program, which sought to promote economic growth in developing countries and thus build political support for America there.[30]

Laurence proposed that the federal government dispatch "a body of leading experts" to Africa to acquire *Strophanthus* plants and seeds and grow them in the Pacific Islands and possibly Hawaii to ensure a US-controlled supply of seeds. Further, Laurence counseled a closed-lipped approach with other governments. "It may be found advisable not to give away our hand until we have assured ourselves of a good supply of the seeds and plants."[31]

Laurence took the midnight train to Washington. Truman's meeting with Laurence on June 28 was scheduled for fifteen minutes but lasted about twenty.[32] The meeting was categorized as "off the record" and so was not listed on Truman's public schedule of meetings for the day.[33] After the meeting, Truman ordered the secretary of agriculture and the head of the Federal Security Agency (the umbrella agency for the US Public Health Service) to act on the problem.[34] "As a result of Bill Lawrence's [*sic*] visit to the President, Mr. Ewing was told to concern himself with this problem," Leonard Scheele, the US surgeon general, wrote about three weeks later.[35] By July 1949 the federal government had organized a botanical expedition to West Africa headed by John T. Baldwin of the College of William and Mary.

By August a US expedition had left for Liberia, and federal officials judged that it was safe to announce the news. Laurence told Turner Catledge, the managing editor of the *Times*, what he had been doing. The *Times*'s employee newsletter claimed that Laurence even dictated the timing of the government's announcement. "He would let the Government make that announcement a day or so after his story had run," the article claimed.[36]

Laurence again rode the train to Washington, where he met with Scheele. Then he took Scheele's government car to the Washington bureau of the *Times*, where he wrote a front-page story for the August 16 issue. He wrote, "The seed of an African plant holds the answer to the prayers of millions for cortisone, the recently synthesized adrenal gland hormone that pours buoyant new life into bodies tortured by arthritis and rheumatic fever, and promises new hope for victims of other chronic ills as well as the mentally afflicted." The article quoted no one and provided no sources but reprised the argument that Laurence made in his memos to Ross and Truman that the vine could be grown to produce vast quantities of seeds. The article said nothing about Laurence's efforts to urge Truman and his administration to send expeditions to Africa to gather seeds that could be used to guarantee American access to the vine. Laurence cannily used the passive voice to imbue his own views with authority, writing that the vine "is expected to become one of the most important plants in the world, promising to serve as the source of a veritable 'elixir of life' for millions of persons the world over."[37]

Later that day, federal officials held their press conference. "This may be to chemistry what the atomic bomb was to physics," Ewing told the reporters about *Strophanthus sarmentosus*, echoing a comparison made by Laurence in his letter to Truman. The *Times* story about the press conference was written not by Laurence but by Bess Furman of the paper's

Washington bureau. The story noted that Laurence had triggered the federal government's interest in *Strophanthus* and quoted Ewing that the government decided to hold the press conference because "so many questions had been raised by the Laurence article"—even though the press conference had been planned before Laurence wrote the article. Curiously, the story was buried inside the *Times* on page 25.[38]

The Associated Press regurgitated the *Times* report: "The New York Times said Tuesday that a rare African plant offers an unlimited source of cortisone, which has shown promise in treatment of arthritis and rheumatic fever."[39] Once again, Laurence scooped the Mayo Clinic's local newspaper; the *Rochester Post-Bulletin*'s front-page article reported that *Strophanthus* seeds could be "the potential source of a new plentiful supply of cortisone" but omitted any mention of Laurence.[40] A reporter from the *Post-Bulletin* did try to get comment from Kendall on Laurence's report. Speaking with a Mayo public affairs official, Kendall dismissed Laurence's plan as "grandiloquent," but Kendall declined to speak with the Rochester paper on the story.[41]

The Rochester paper wasn't the only media outlet to ignore Laurence while writing about the search for the African vine. *Newsweek* omitted his role in the quest for the African plant, noting only that "a report [had] leaked out" that the Public Health Service was leading the search for sources of the plant.[42] *Time* did describe Laurence's quest but backhandedly said that as a result "hopes were briskly and perhaps brashly fanned for a short cut in production" of cortisone.[43] The newsweekly quoted Mayo's Kendall as saying that other sources of cortisone would probably be developed before enough *Strophanthus sarmentosus* plants could be grown to become a viable source of the hormone. The *Los Angeles Times* ran an AP story that focused on the US government's quest to find *Strophanthus* seeds; the story did not mention Laurence or the *Times*.[44] The *Chicago Daily Tribune* ran a six-paragraph item, deep inside the paper, that described the vine's potential but never cited Laurence, the *Times*, or any other source for its reporting.[45] United Press also downplayed Laurence's behind-the-scenes role. The news service reported the efforts by the US government to find the plant but did not mention Laurence until the story's sixth paragraph.[46]

After the first wave of stories, other science journalists around the country largely ignored the *Strophanthus* story, while the *Times* developed something of a cottage industry in the topic. Two days after Laurence's story about the vine, the paper ran an unbylined article focusing on the fact that the New York Botanical Garden was displaying a *Strophanthus* plant in its main conservatory.[47] Three days after that, Laurence published a story focusing on the potential of cortisone and the difficulty in obtaining it. Laurence wrote self-servingly: "It was therefore most encouraging news when it was revealed in this newspaper last Tuesday that the seed of an African plant, known as Strophanthus sarmentosus, produces a substance that could serve as a substitute for the animal product, and that President Truman had sent an expedition to Liberia to obtain seeds of the plant." That day's edition of the paper also ran a drawing of the *Strophanthus* plant, under the headline "The 'elixir of life.'"[48]

Soon, clouds began to gather. Russell E. Marker, a researcher at the Treemond Pharmaceutical Company, identified a Mexican yam named *Discorea mexicana* as another possible source of an ingredient for manufacturing cortisone. The yam was seen as more practical than the vine as a source of cortisone. Alton Blakeslee, the AP reporter whom Laurence had deserted in Detroit in order to get to the Mayo meeting in April, quoted a Treemond official as saying that the yam could be used to produce cortisone "in the near future."[49] *Time* described the African vine as "over-trumpeted" and said that the yam's discovery was "another slender ray of hope" for arthritics that had the advantage of requiring "no costly task force (like the one sent to Africa to gather Strophanthus seeds)."[50]

The yam's discovery handed ammunition to Kaempffert, Laurence's co-worker and rival at the *Times*, who sought to undercut the *Strophanthus* story. Kaempffert reported the yam's discovery twice in a single day's edition. One article, in his Science in Review column, reported the details. In an analytical article deeper in the paper, Kaempffert declared that it was "likely" that pharmaceutical companies would favor the yam as a source of cortisone. Offering faint praise for *Strophanthus*, Kaempffert wrote that the uncertainties surrounding the vine's utility as a source of cortisone were such that the *Strophanthus* expedition "is not necessarily wasting time and money."[51]

Meanwhile, the paper's management embraced Laurence's quest for the vine. Two days after Laurence's story ran, Orvil Dryfoos, an assistant to Sulzberger who later would become the paper's publisher, wrote

Sulzberger to say that "Laurence can truly be said to be the real discoverer" of *Strophanthus* as a source of cortisone. "The German document [that Laurence found] was the missing link in the research work," he wrote Sulzberger.[52] Almost two weeks after Laurence's story ran, the *Times* ran a highly unusual editorial praising him by name. "Mr. Laurence deserves great credit for his role as catalyst in the progress made to date toward the solution of this problem. In his search for news, he has brought to light invaluable facts for further research."[53] In a private note to Laurence two days after the editorial ran, Sulzberger called it a "beautiful story." Laurence replied that the editorial was "the highest honor that has come to me so far."[54]

The only objection that *Times* managers raised about the story regarded Laurence's efforts to hog credit for his work. *Times Talk*, the in-house employee newsletter at the *Times*, described Laurence's work on getting information on the seeds: "All this detective work had been done in his off-time."[55] Catledge wrote to Sulzberger to complain that Laurence had stated that he pursued the project "on his own time and at his own expense. He did not have to do either." Catledge wrote, "He would have been given all the time he wanted, with no questions asked, had he simply said he was working on something important and needed to be relieved of other duties."[56]

Sulzberger too was displeased about this side of the story. On August 29—the same day he wrote Laurence to praise his "beautiful story"—Sulzberger wrote the author of the *Times Talk* article to complain about that aspect of her story about Laurence. "He could have done it on office time and the emphasis which it is given seems to indicate that that would not have been the case. That's a little unfair to us, I think."[57] She defended herself by return memorandum, stating that she had meant no criticism of the paper's managers. "The only impression I tried to convey in the story about Bill Laurence was that this was a reporter so carried away by his zeal that he chose to use his off-time and some of his own money to get at the story."[58]

No journalist, inside the *Times* or outside it, seemed to have any concerns that Laurence had morphed from a neutral observer into someone advocating a particular project by the federal government. Bill Hollenbeck, a CBS radio commentator, praised Laurence's "detective-diplomatic achievement."[59] A barometer of attitudes throughout American journalism, the trade industry magazine *Editor and Publisher* detailed Laurence's escapade without any tone of question or rebuke.[60]

Meanwhile, some key researchers were skeptical that Laurence had pinpointed a source for plentiful cortisone. At one scientific meeting, Kendall was dismissive of both the vine and yams as sources for cortisone, while Fieser said both likely would yield small amounts. In all, the participants at the meeting were about equally divided over whether plant sources would be better than starting from scratch. The *Times* put the best possible face on its coverage of the meeting, with the headline, "Some Experts See Hope in Plant Source."[61]

A month later, in a front-page story, Laurence reported on the use of sex hormones to reduce symptoms in arthritis patients. Halfway through the story, Laurence added a diversion arguing that the sex hormones were only a stopgap until cortisone became available in large quantities. "The best hope for the present to obtain a more abundant source of starting material for cortisone lies in the seed of an African plant, *Strophanthus sarmentosus*," he wrote.[62]

In late September another source of cortisone emerged: soybeans. The *Times* ran a five-paragraph version of the AP story inside the paper. Even though the science journalists at the *Times* never produced any stories on the soybean research at that time or later, soybeans eventually would become the raw material used in most cortisone manufacturing.[63] The *Times* greatly publicized *Strophanthus* but missed the actual solution for the cortisone problem.

Rival hormones were surfacing—most importantly adrenocorticotrophic hormone, or ACTH, which also showed promise in treating arthritis. Journalists sought to describe the relative merits, challenges, and modes of action of cortisone and ACTH.[64] ACTH was derived from the pituitary glands of pigs and had been developed by a division of Armour and Co., a meatpacking company. The *Chicago Daily Tribune* took a particular shine to ACTH, perhaps because of the role played by the meatpacking company. The paper's coverage emphasized that ACTH was less scarce and easier to synthesize than cortisone.[65]

It was a time of excitement about hormones among both medical researchers and the science journalists who covered them. On October 1 the AP had two different stories about the use of hormones. One, datelined Boston, reported more extraordinary results of treatment with cortisone; a second, datelined New York, reported a journal paper that showed that sex hormones provided temporary relief from arthritis. Later in the month, researchers told the American Chemical Society that injections of cortisone and ACTH had shrunk some tumors. At the *Los Angeles Times*, science writer William S. Barton also preferred ACTH. "Perhaps cortisone . . . never will be needed" because of ACTH's efficacy, he wrote.[66] A few days after Barton's story appeared, Kaempffert was again deflating the *Strophanthus* story. With more than a half million species of plants on Earth, "it would be astonishing that of this vast number only one contains enough of the right raw material" to make cortisone, he wrote.[67]

As 1950 dawned, skepticism was growing about both the use of cortisone in treating arthritis and about the African vine as a source of cortisone. In January Scheele traveled to Capitol Hill for a hearing before the powerful House Appropriations Committee about the proposed budget for the Public Health Service. Lawmakers clamored for quicker access to cortisone. "I have been besieged by people who are seeking help as a result of reading these articles that appeared in the press and in magazines," said Frank B. Keefe, a Republican congressman from southeastern Wisconsin. He added that the pressure was not just from constituents but often was more personal: "I know a number of the wives of Members of Congress who are afflicted with arthritis, for instance. Hope has been held out to them and yet . . . the results are not attainable," he told Scheele.[68] The surgeon general counseled patience. "Those articles, I am afraid, as far as offering immediate hope is concerned, were a bit premature," he said. "On the other hand, I think the claims made in those articles will eventually be proven to be correct."[69]

At the same time, researchers were realizing that Laurence's beloved *Strophanthus* vine would not prove to be a practical source of cortisone. Just after New Year's Day in 1950, the *Times* ran an unbylined story, datelined Liberia, quoting an official from the US Public Health Service as predicting that producing cortisone from *Strophanthus* would take at least two decades.[70] Six days later, Kaempffert reprised the statements, noting that it was "not cheerful news," but arguing that the Liberian expedition had a mission "which is only botanical," not pharmaceutical.[71] About five months later, the paper ran another unbylined story noting the return of

Upjohn's expedition to the United States and calling *Strophanthus* "the 'Cinderella' of the botanical and medicinal worlds."[72] Three days later, Kaempffert repeated the news of the return of the botanists but ended, "It still has to be established that Strophanthus sarmentosus is a richer source of the raw material from which cortisone can be derived than one of the 400 Mexican plants."[73]

In February 1950 Kaempffert wrote an article that was deeply skeptical of the safety of cortisone and ACTH, reporting that the Arthritis and Rheumatism Foundation was downplaying concerns about the hormones' risk, while the American Medical Association's Council on Pharmacy simultaneously was opposing use of the drugs because of their side effects. "This department doubts if any brain surgeon who cares about his reputation would approve the lifelong use of cortisone and ACTH," Kaempffert wrote. "Either the council or the foundation is wrong, and this department would like to know which."[74]

The article enraged the Mayo Clinic's Hench, who blamed Kaempffert's coverage on his long-standing rivalry with Laurence. Hench wrote two family members that he had been told by Rush that "Kaempffert and Bill Laurence had been feuding for years and were on opposite sides of many arguments. . . . Whatever Laurence favors Kaempffert tries to discredit." Moreover, Kaempffert based his reporting on press releases and advertisements rather than published journal articles and as a result was "raising false hopes" for alternatives to the hormones, Hench wrote.[75]

Others felt Laurence was the one exaggerating the chances for treatment. "My mother suffered dreadfully from arthritis, and Mr. Laurence gave her and other arthritics false hopes at too early a stage of the research," Anne Louise Davis, of New Jersey, wrote Sulzberger in 1954. "He reports the constructive, hopeful side of medical stories, but doesn't report the negative side."[76]

Laurence remained a true believer, touting the vine in the January 1950 issue of *Life and Health*, published by the Seventh-day Adventist Church. Given the scarcity of ox bile, "it was therefore most encouraging news when it was revealed in the New York Times" that the vine's seeds could be a source of cortisone, Laurence wrote self-servingly.[77] In a lengthy feature article in the *Saturday Evening Post* in April 1950, Laurence cast the discovery of cortisone as a turning point in efforts to reduce death from disease and thus extend life expectancies. "We now stand at the doorway of an era in which we shall, within limits, be able to turn back the clock of biologic time," Laurence declared.[78]

That same month, Laurence received one of the first Albert and Mary Lasker Awards for medical journalism for his reporting on cortisone and ACTH. In presenting the award to Laurence, Scheele praised Laurence for his reporting on the 1949 Mayo meeting. The surgeon general said that Laurence "also gave scientists a new lead on the possibilities for mass production of the new drug."[79] When Kendall and Hench were named to receive the Nobel Prize in Medicine or Physiology in October 1950, Laurence dispatched a congratulatory telegram to Kendall that made the news as much about him, writing, "The award of the Nobel Prize to you for your epoch-making work came as no surprise to me who knew it since April 20, 1949."[80]

In all, five different expeditions trekked to Africa in search of seeds of *Strophanthus sarmentosus*. The pharmaceutical companies Merck, Upjohn, and Ciba each sent an expedition; the US and French governments also sent one apiece. Growing the vine proved impractical, not least because the seed-containing fruits were out of easy reach, growing at the top of the vines, 100 feet or more above the ground.[81] In early 1950 Joseph Monachino, a well-known botanist at the New York Botanical Garden, suggested that the seeds that had so excited Laurence had been mistakenly identified and probably had come from another *Strophanthus* species and thus were not relevant to the search for a cortisone source.[82] A team of Swiss scientists concluded that seeds that had been definitively identified as coming from *S. sarmentosus* did not include appreciable amounts of the chemical that could be used to make cortisone. "All the elaborate steps taken to procure large seed samples and to commence cultivation have apparently been directed at the wrong species!" Monachino exclaimed. The scientist listed other challenges with the vine besides identifying the specific species that was needed, including the difficulty in growing the plants and the death of one of the most knowledgeable researchers in the field. "The future of Strophanthus appears dark, indeed."[83]

He was correct. In February 1951 Howard W. Blakeslee, the Associated Press's science editor, reported that Merck had ramped up production of cortisone even without using the *Strophanthus* seeds or Mexican yams. "Today your doctor can buy cortisone to treat you, and that is the start of a new era in medicine," Blakeslee wrote.[84] In its annual report, issued in April 1951, the Federal Security Agency was pessimistic about the *Strophanthus* seeds, saying that results of preliminary tests were "discouraging. . . . Strophanthus probably will not be a practical source of cortisone, but the search for a suitable plant source will be continued."[85] The *Times*, which previously was willing to expound on the bright promise of the *Strophanthus* vine at great length, limited the discouraging news about the vine to paragraphs 16 and 17 of a story focused on federal expenditures on health research.[86] After that story, the word *Strophanthus* has never reappeared in the *Times*.

By 1952 Upjohn had devised a way to use the mold *Rhizopus nigricans* to make cortisone, and that opened up the supply lines.[87] By 1955 researchers had dismissed the vine as a source of cortisone.[88] "The search for the lost strophanthus proved an expensive waste of time," a history of the Upjohn Company concluded.[89]

Laurence never lost faith in the vine—or his analysis of its potential. "I am convinced that there is a lot yet to be found," he told a book author, apparently in the early 1960s. "I know all about the difficulties involved, but I still think they should pursue it. I am, naturally, disappointed that they dropped it."[90]

10

The Hell Bomb

During the wartime Manhattan Project, the physicist Edward Teller had spent much of his time considering how to design a weapon that was based on nuclear fusion (a hydrogen bomb, or H-bomb) rather than the fission process used in the Hiroshima and Nagasaki weapons (atomic bombs, or A-bombs). But after the war ended, American nuclear superiority eroded interest in developing the Super bomb, or Super, as Teller had dubbed the fusion weapon, and work slowed.

The Soviet Union's first atomic test on August 29, 1949, breathed new urgency into developing an American fusion bomb, to counter the perceived threat of an atomic-armed Soviet Union. Controversy raged among US policy makers, weapons designers, and the military. But Laurence seemed to have moved on; in late 1949 he wrote nothing about the hydrogen bomb debate roiling the top levels of government, instead covering photosynthesis, medicine, and hormones.

Finally, the columnists Joseph and Stewart Alsop revealed the existence of the controversy within the Truman administration over whether to build a hydrogen bomb, which would be far more powerful and fearsome than the atomic bomb. In a series of columns starting January 2, 1950, the Alsops intimated that the H-bomb, which has been called the Hell Bomb because of its massive destructiveness, could require an industrial effort even more enormous than the Manhattan Project.[1]

At first, Laurence remained absent from the debate over the H-bomb, and the *Times* as a whole was surprisingly slow to cover it. The *Times* did not subscribe to the Alsops' column, so the newspaper's first reference to the possible H-bomb project was in a United Press story that was tucked onto page 26 of the *Times* almost a week after their revelation. Others at the *Times* then started chasing the story; on January 17 James Reston had a front-page *Times* story that Truman was being urged to seek an arms control agreement with the Soviet Union before deciding to proceed with the H-bomb project.[2] The next day, Laurence used the term "hydrogen bomb" in print for the first time, but he downplayed the import of the project, writing that much of the data needed for an H-bomb already was in the scientific record. Laurence also lowballed the likely power of the hydrogen bomb. "The first models, when and if constructed, would have no more than ten times the explosive power of the present models," Laurence claimed, quite mistakenly.[3] Eight days later, in response to press reports that some weapons scientists had reservations about developing a hydrogen bomb, Laurence wrote another story predicting that weapons scientists would not refuse to work on the project.[4] (But within ten days, Laurence contradicted himself with a front-page story reporting that twelve major physicists were collectively calling on the United States to pledge not to be the first to use a hydrogen bomb.)[5] Meanwhile, other journalists were busy speculating on technical details of the bomb, a subject Laurence did not touch. Although Teller and the physicist Hans Bethe had informed Laurence during the war that an H-bomb was possible, classification rules had barred Laurence from revealing the work on the hydrogen bomb that had been conducted through the Manhattan Project. So when the Alsops revealed the prospective H-bomb project, Laurence felt that the security rules had unfairly forced him to miss the story.[6] *Vogue* reported, "With his knowledge of atomic power, Laurence sometimes feels he is 'sitting on a powder keg and constantly worrying about letting the cat out of the bag'; that he must bend over backwards to keep the public in the right amount of darkness, a major contradiction in a news reporter's job."[7]

When Truman announced his decision to proceed with the H-bomb project on January 31, 1950, the main story in the *Times* was written not by Laurence but by Anthony Leviero, a reporter in the Washington bureau who within six years would develop into a public critic of Laurence's work. Laurence instead wrote a front-page sidebar that discussed the technology of the planned weapon. Unlike most other reporters, he emphasized the role of tritium, a radioactive isotope of hydrogen, which would add to the bomb's explosive power. In fact, Laurence cheekily corrected Truman's choice of terminology in calling the device a hydrogen bomb, writing that "it is not a hydrogen bomb at all in the true scientific meaning of the term." Rather, Laurence claimed, the device would be more properly called a triton bomb because of its reliance on tritium.[8] So powerful was Laurence's influence that newspapers throughout the nation and as far away as Australia initially ran stories citing Laurence arguing that the device should be called the triton bomb.[9] But the term did not stick; today no one calls it that. Even the *Times*, famously persnickety about language, used the arcane term only four more times, all within three weeks of Laurence's story, before abandoning it.

Soon Laurence moved to a more sensational approach to covering the hydrogen bomb, emphasizing its destructive power and environmental impact. In a speech in mid-February to an audience of four hundred at the Enoch Pratt Free Library in Baltimore, Laurence called the bomb a "hideous thing threatening the Earth itself." Rather than downplaying the bomb's power as he had previously done, now he held that the H-bomb would be incredibly more powerful than the A-bomb, so much so that its very existence might deter any use of such weapons. He predicted that the H-bomb could be "several million times more powerful" than an atomic bomb and said that it would release so much long-lived radiation that "bombed cities will be uninhabitable for possibly thousands of years."[10]

About the same time, concern about radioactive fallout from the H-bomb jumped internationally, thanks to a radio program in Chicago called the *University of Chicago Round Table* in which scholars, public officials, and intellectuals debated the issues of the day. In its episode for February 26, the *Round Table* featured four atomic scientists talking about the hydrogen bomb. The physicist Leo Szilard, who held a patent on the chain reaction that was central to atomic power, horrified the other panelists and listeners by suggesting that a single H-bomb could kill all life on Earth if the bomb were encased in material that would absorb neutrons from the explosion and become radioactive in the split second between the bomb's detonation and the dispersal of the material by the force of the explosion. That process would generate fine radioactive dust that would spread throughout the atmosphere, rain down on Earth, and remain lethally radioactive for years or decades. Szilard warned listeners that it would be "very easy to rig an H-bomb on purpose so that it should produce very dangerous radioactivity."[11] News coverage appeared nationwide, with *Time* headlining its article "Hydrogen Hysteria."[12]

Szilard did not specify the material to be used in the casing, but Laurence then wrote a front-page story for the *Times* identifying cobalt as the bomb casing that could generate enough radioactive fallout to kill all life on Earth. Over the next several years, Laurence would repeatedly raise the specter of the cobalt bomb. For example, in June 1950 he predicted in the *Saturday Evening Post*, "Thus a large area could be made unfit for human habitation for months or years."[13]

Controversies such as the one over the cobalt bomb pushed public fears and curiosity about atomic energy to new heights. Trying to satisfy that hunger, radio producer Fred W. Friendly in 1950 got an assignment from the National Broadcasting Company to produce a radio documentary about nuclear technology. Friendly turned to Laurence, whose overwhelming journalistic profile on atomic energy positioned him as a technical expert. In the summer of 1950, NBC agreed to pay Laurence $1,400 for him to be interviewed for about fourteen hours, to allow NBC to adapt *Dawn over Zero* and his then-forthcoming *Saturday Evening Post* article, and to allow his voice to be recorded. Laurence also presented "special lectures" for the staff preparing the show.[14] Friendly wrote the script for the series based on information he drew from *Dawn over Zero* and from the interviews with Laurence conducted by his staff in Laurence's apartment between late May and mid-June 1950.[15]

The premise for the radio series, titled *The Quick and the Dead*, was an extended conversation between Laurence and Bob Hope about atomic energy. Hope was portrayed as an Everyman who was curious and a bit confused, and Laurence patiently answered Hope's questions. "I happened to look at my income tax form, and I realized I was a very big taxpayer, and I knew an awful big part of that tax was going toward atomic energy plants of which I knew less than nothing," Hope said in the introduction

Figure 10.1. A publicity photo for *The Quick and the Dead*, a 1950 NBC radio series featuring Laurence and Bob Hope that explored the history and uses of atomic power. (Library of Congress)

to the first episode. "So one day I figured as long as I'm financing this project, I ought to kind of check up on things." In reply, Laurence tells Hope the story of the atom bomb. "It wasn't my story," Laurence says. "It was mankind's story and perhaps the biggest news story of the 20th century. I was privileged to be tapped on the shoulder by the Army and given the assignment that every newspaperman dreams of."[16]

Their dialogue was interspersed with dramatizations of key events from the history of atomic technology, with some participants reenacting their own voices. While recording *The Quick and the Dead*, Hope and Laurence never were physically present in the same recording studio. Hope recorded his side of the dialogue at a studio in Hollywood, while Laurence recorded his portion in New York. Indeed, all the voice actors in the series recorded their parts separately from one another, for subsequent assembly into a finished program. In an era in which plastic-based audiotape was only beginning to come into broadcasting use, this editing technique was an innovation.

The first episode related Laurence's visit to view the Trinity test. Along the way, the listener learned about the basic description of an atom, the insights of Lise Meitner and Albert Einstein, and the first nuclear reactor underneath the Stagg Field stands at the University of Chicago. The second show in the series focused on the construction of the Oak Ridge plant and the Hiroshima bombing. The third discussed the hydrogen bomb, and the fourth covered the peacetime uses of atomic energy.

Broadcast over four weeks in July 1950, *The Quick and the Dead* was very well received. A reviewer for *Billboard* magazine said that the first installment was "exacting, exciting theater" that "demonstrated radio's great possibilities as a contributor to America's education and solidarity as a nation." But the reviewer also noted Laurence as "speaking with a heavy accent."[17] The reviewer for *Variety* also took exception to including Laurence's voice in the program. "A well-chosen professional surely could have heightened the value of the scientific answers that were handled by the N.Y. Times science expert."[18]

The series was so popular that NBC made the then-unprecedented decision to rebroadcast a condensed version. "The tremendous reaction to the series, by the public, is undoubtedly caused by the present global unrest and the breaking out of war in Korea," a radio industry publication observed.[19]

Afterward, Friendly praised Laurence's participation. "He worked much harder than anyone thought he would have to, and was an invaluable asset not only in presenting the program, but in getting the cooperation of scientists." Friendly asked that NBC give Laurence a television set; the gesture "would probably be a very inexpensive way of getting repeat rights for this series."[20]

The network decided to release *The Quick and the Dead* on phonograph records. But the record set drew less favorable reviews than the radio broadcasts. The reviewer for the *Saturday Review* called it "slow moving; Laurence's accented oratory is sincere but clumsy; Hope's hopisms forced, his role a bit foolish throughout. Splendid exposition of vital civilian information, but labored."[21]

The public concern about the risks of atomic war, particularly fallout, also fueled the development of documentary and training films about civil defense, showing viewers how to protect themselves if they were caught in an atomic attack. Always on the prowl for side work, Laurence agreed to work on one such film, *Pattern for Survival,* released in December 1950, which offered a platform for him to promote the viewpoint that nuclear war would be survivable if civilians could shield themselves from the blast. Implicitly this also argued that long-lived radioactivity was not a major concern, a position that he had taken about fission bombs since their first use. Laurence served as a technical adviser to the film.[22]

The twenty-minute film opens with a tour through a museum display of weapons used throughout history. An off-screen narrator tells the viewer that weapons have generated fear throughout the ages. As a mushroom cloud rises, the narrator says that today's people must wrestle with a greater fear—and then the movie fades to a cigarette-smoking Laurence, set in his "office," replete with cluttered desk, desk lamp, and typewriter. The narrator reminds the audience of Laurence's Pulitzer Prizes and two books and says that he is "one of the persons most qualified to give us his views for an approach to a method of defense against atomic warfare."[23]

Laurence then gazes into the camera and says,

> I saw with my own eyes the power of the exploding atom. I witnessed the end of an era and the beginning of a new one. I saw a world blow up in a burst of cosmic fire and a new one born from its ashes. I saw a great city disappear in a mushroom cloud in a fraction of a millionth of a second. Right then I knew that here at last man had a weapon that could destroy civilization. The atomic bomb is the most powerful physical destructive weapon history has ever known, but it is above all a psychological terror weapon. Does this mean that we are helpless against an atomic attack? Most certainly not.

As the camera zooms in, Laurence continues, "There is definitely a defense against the atomic bomb provided we faithfully carry out a planned method of defense." With head cocked and brow furrowed, Laurence intones, "Note very carefully what is to follow, for what you are about to see and hear is your pattern for survival."

Next, Laurence hands the film off to the journalist Chet Huntley, then a CBS radio reporter, who explains in detail the duck-and-cover strategy that the film is promoting: to protect themselves from the bright blast of an atomic bomb, people should wear light-colored clothes, duck behind a covering area, and cover their eyes with their arms. Although the area closest to ground zero is likely to be wholly destroyed, with care people in a surrounding area could survive, Huntley tells them. The movie dismisses the long-term risk of radiation from an airburst bomb, reserving that threat for a bomb detonated underwater, which would force contaminated water into the air, whence it would rain on people in surrounding areas. The movie ends by encouraging viewers to support the emergency workers who today would be called first responders. It never returns to Laurence.

Cornell Film Company, which produced *Pattern for Survival*, capitalized on Laurence's role in the film. In one letter to community leaders, the company's president, J. Milton Salzburg, wrote, "The picture features Mr. William L. Laurence, noted science writer for the New York Times, and two time Pulitzer Prize winner for his books on the atom bomb, and the only newspaper man who was assigned officially to cover the entire Atom Bomb Project." A print advertisement for the film, designed for use in trade journals, announced "Pattern for Survival featuring William L. Laurence, World-Famous Scientific Editor of the New York Times." The ad could not have summarized Laurence's own views more succinctly: "Fight fear with knowledge . . . and you may not have to fight panic later."[24] The *Times*'s own article about the film highlighted Laurence's role and quoted him as saying in the film that defense against the bomb is possible "if we do not permit fear to overcome our power to reason." The unbylined article describes the film's plot for several paragraphs.[25]

The premise of the film—a massive atomic attack on civilian centers in the United States—was unsettling to some. "Audiences may be jolted by the subject," one reviewer warned.[26] Nevertheless, *Pattern for Survival* was widely characterized as informational rather than propagandistic. *Saturday Review* said *Pattern for Survival* "gives factual information on atomic radiation and suggestions about personal preparation and protection against atomic attack." The "major criticism" of the film is that it is not sponsored by the US government.[27]

Projects such as *Pattern for Survival* ate up a good portion of Laurence's time, and despite the free publicity that the film gave the *Times*, the relationship between Laurence and the *Times* was fraying. His editors had good cause for dissatisfaction: after Laurence published 108 bylined articles in the *Times* in 1948—more than he ever had published in a single year—the count then turned downward, to ninety-nine articles in 1950 and only fifty-two articles in the *Times* by 1954. Meanwhile, his outside work mushroomed. During the 1950s he wrote at least fifteen original articles for popular magazines and had another seventeen articles republished in magazines.

Simultaneously Laurence complained about being underpaid by the *Times*. This flared up in August 1950, when Laurence wrote a series of articles based on an AEC book describing the effects of the use of atomic weapons. Even before all installments in the series had been published, the publisher McGraw Hill approached Laurence to reprint the series, for which Laurence believed he would have earned at least $25,000. But the *Times* refused permission for the project, instead reprinting the articles itself in a booklet sold for ten cents.[28] Laurence resented that decision, and he sought more outside projects to supplement his *Times* salary.

Laurence closed out 1950 with the December release of his next book, *The Hell Bomb*, which aimed to explain the hydrogen bomb to a nontechnical audience. Just days after Truman had decided to proceed with development of the H-bomb, Knopf had paid a $2,000 advance to Laurence, and in return he agreed to produce a manuscript about the H-bomb of approximately the same length as John Hersey's *Hiroshima*.[29] But the editors learned again that Laurence could wreak havoc with deadlines, and keeping his writing short was not exactly one of his specialties. On June 23 Laurence's agent, Alan Collins of the Curtis Brown agency, assured Knopf that Laurence's manuscript would be in their hands around July 4. But three days later, Collins had to eat his words. "I don't understand why authors won't be honest even with their own agents," he wrote to Knopf. "Laurence told me a week ago that all his work on the book was done and that all that remained was clearance by censorship. Now it seems that only about 20,000 words of the book are in the hands of censorship and the rest remain to be written when he returns from California on July fourth." The manuscript was submitted on July 31, with Laurence explaining that the delay was needed to bring the text up to date.[30]

The Hell Bomb contained five chapters and described some of the technical problems in designing a hydrogen bomb, discussed the cobalt bomb, and complained that the bomb could have been ready already if the

Truman administration had not sidelined the project. In the second chapter, titled "The Real Secret of the Hydrogen Bomb," Laurence concluded that a hydrogen bomb had to incorporate both deuterium and tritium, not solely one or the other isotope of hydrogen. He estimated that a workable bomb might include 150 grams of tritium and 100 grams of deuterium. Laurence analyzed and rejected the notion, then popular among some scientists, that the United States should unilaterally renounce the use of the hydrogen bomb, although he did support forswearing the use of a cobalt bomb or other "rigged" hydrogen bombs because the resulting radioactive fallout would have no military purpose. In the book's fourth chapter, Laurence argued that the just-started Korean War demonstrated the need for a stockpile of hydrogen bombs to deter conventional attack. The book concluded with a primer on atomic energy and an appendix on issues about international control of atomic weapons.

After reviewing the manuscript, one editor at Knopf compiled a list of "overused words" in Laurence's text: *basic* and *basis*, *simple arithmetic*, *20,000 tons of TNT*, and *fundamental*. The editor suggested that Laurence should eliminate quotes, shorten sentences, reduce the use of adjectives, and watch for misuse of certain words. Two weeks later, editor Harold Strauss sent Laurence the edited manuscript. Laurence delayed returning the edited manuscript until August 28 and then through his agent immediately requested his advance. But complaining that "Laurence's working habits are so strange" and citing the fact that Laurence had submitted new text that still needed to be edited, Strauss withheld the advance until satisfied with the final manuscript on September 6. During the editing process, Laurence made so many handwritten changes in the typewritten manuscript that Strauss fretted the printer would charge the publisher extra.[31]

Anxious for a wide audience for the book, Laurence nagged Knopf to offer it to the Book-of-the-Month Club, which had distributed *Hiroshima* to club members for free. The writer Fletcher Pratt did recommend it adopt *The Hell Bomb*, calling it "an extremely careful, comprehensible and vividly interesting report on the Hydrogen Bomb." But retired army brigadier general Donald Armstrong, a business executive and author, advised the club against accepting the book, and his judgment apparently held sway, as the book club declined.[32]

Once the book landed in stores in early December, it sold well, at the rate of more than a thousand copies each week, and Laurence hit the lecture circuit to promote it. He said at the University of Michigan, "We cannot leave any stone untouched to overcome the superiority of manpower which the U.S.S.R. and Soviet China have over the United States. We are forced to manufacture the hydrogen bomb."[33] In January 1951 Laurence assured a Newspaper Advertising Executives Association meeting that the Soviet Union's "jalopy" atomic weapons did not measure up to the United States' "super 1951 model."[34] Three months later, Laurence exhorted an audience of advertisers that they should "defrighten" the American people so that US leaders could "call the bluff" of the Soviet Union, which was lagging badly behind the United States in the ability to produce atomic weapons.[35] But as an immigrant, Laurence did bring out xenophobia in some, such as a Brooklyn man who wrote to the *Times* to complain, "It seems to me that Laurence speaks with a foreign accent—and it occurs to me that this man is working for the destruction of America. Do you have any comment?"[36] The files of the *Times* show no record of a reply.

Laurence's friend Howard W. Blakeslee, the AP science editor, touted *The Hell Bomb* in a feature story that was timed to coincide with the book's release. The article started, "Hydrogen bombs are considered possible by science writer William L. Laurence—in fact, he tells a new way by which they may be made successfully," a reference to Laurence's proposal to mix deuterium and tritium. The article showed the extent to which Laurence was regarded, at least by fellow journalists, as a technical expert in weapon design.[37] But the *New Republic* complained that Laurence "has a distinct tendency to throb and to exaggerate. Everything is 'enormous,' or 'major,' or 'top,' or is described by some other banality of grandiloquence."[38] The *New Yorker* called it "a book partly describing and partly guessing at the nature of the hydrogen-fusion bomb."[39] *Life* was of two minds about the book. "When Laurence reports on what the U.S. has done and is doing, his account is generally sound and informative; his deductions from this information, though only assumptions, are provocative; but when he gets into the field of military strategy and of estimates of enemy strength, his logic is considerably less impressive."[40]

Many technical experts challenged the book's accuracy. George Paget Thomson, the British physicist and Nobel laureate, said that Laurence's position on the amount of tritium needed for a hydrogen bomb amounted to little more than a guess.[41] William A. Higinbotham, a physicist who headed the Federation of American Scientists, wrote in a review in the *Times* that Laurence made questionable assumptions. "We cannot take the book on faith, and it certainly does not prove its case."[42] A reviewer for a Caltech magazine challenged Laurence's conclusions. "He is, in fact, alone in many of them; few fully-informed scientists would agree with him. Unfortunately, few of the laymen who read his book will be well-enough informed to know that."[43]

The most detailed critique came from Robert F. Bacher, a physicist at Caltech, who coincidentally was quoted at length by Laurence in the book's introduction. *Reader's Digest* purchased rights to publish a condensed version of the book and planned to publish it in April 1951, but editors there got cold feet. In January of that year Paul Palmer, a senior editor at *Reader's Digest*, recruited Bacher to advise the magazine about "the accuracy of the facts presented in this piece, as well as the conclusions drawn by the author." In a follow-up letter, Palmer confided that he was most concerned about the validity of Laurence's conclusions that the hydrogen bomb could obliterate life on Earth. "Of course we want to tell our readers the truth—but how difficult it is for the layman to be sure of *what* is the truth in such matters!" Palmer wrote. If Laurence's claims were unwarranted, Palmer wrote, "we would be amiss in unnecessarily alarming the American public."[44]

Bacher panned the book. "There are a good many errors in Mr. Laurence's book, some of them small and some of them serious. They can not be pointed out in any detail without violating security. Mr. Laurence is in no position to know that he has made some of these errors since, I believe, he does not have access to current work on the subject. He can hardly be held responsible, but the difficulty is that many people will consider him an authority. . . . The impact of Laurence's book is that of a scare story."[45]

Bacher's critique did the trick; the magazine did not run the Laurence condensation. His agent told Laurence that the *Digest*'s editors felt "that the entire question of the H-bomb was not as imminent as implied in your book." Nevertheless, *Catholic Digest* did publish an excerpt.[46]

Laurence also used *The Hell Bomb* to criticize the German theoretical physicist Klaus Fuchs, whose spying on the Manhattan Project for the Soviet Union had come to light in January. Laurence maintained he had never met Fuchs, even though the two were at Compania Hill for the Trinity test in 1945.[47] This portrayal of Fuchs was something of a change for Laurence, who had treated him with some evenhandedness in a *Times* story after his confession in 1950, referring to "aid alleged to have been given to Russia by Dr. Klaus Fuchs." In that story Laurence quoted the physicist Frederick Seitz as saying that if Fuchs had aided the Soviet Union in developing an atomic bomb, "it no doubt was a considerable help, but was by no means the determining factor."[48]

But in *The Hell Bomb* Laurence was much harder on Fuchs, writing that Fuchs might have put the Soviet Union ahead of the United States in development of a hydrogen bomb. Laurence wrote that physicists at Los Alamos were saying that Fuchs enabled the Soviet Union to slash at least a year off the schedule for developing their hydrogen bomb. Laurence's own estimate was more severe. "It is my own conviction that the information he gave the Russians made it possible for their scientists to attain their goal at least three—and possibly as much as ten—years sooner than they could have done it on their own," Laurence wrote, citing the fact that Truman had announced his decision to proceed with development of the H-bomb just days after Fuchs's confession. "This confession made it evident that the Russians were without doubt already at work on the hydrogen bomb and had probably been working on it uninterruptedly since 1945. The tragic prospect is that instead of the Russians catching up with us, it is we who may have to catch up with them."[49] Laurence repeated this analysis in the *Saturday Evening Post* in June 1950, but he did not share this claim with readers of the *Times* until after the Soviet Union announced in August 1953 that it had developed a hydrogen bomb.[50]

The truth about the impact of Fuchs's spying remains unclear even today. Another veteran atomic reporter, Bob Considine, also maintained that Fuchs had saved the Soviet Union a decade in its development of a hydrogen bomb, but the historian Ferenc Morton Szasz termed that "a minority view," with most observers pegging the advantage at only one to two years.[51] Waldemar Kaempffert, Laurence's *Times* colleague who often was as much a competitor, estimated that Fuchs saved the Soviet Union eighteen months to two years.[52]

The Hell Bomb also bolstered the reputation of Edward Teller by inflating Teller's role in the H-bomb project. Teller—who later would play a controversial role in the revocation of the security clearance of J. Robert Oppenheimer and, decades later still, in the launch of the Strategic Defense Initiative by President Ronald Reagan—was an important news source for Laurence, who mentioned Teller by name in thirty-nine articles bearing his byline in the *Times*. Laurence's landmark September 1945 article about the Manhattan Project depicted Teller at the Trinity test warning observers to guard against sunburn from the fireball.

Only after the publication of *The Hell Bomb* did other journalists start depicting Teller as in the driver's seat for the H-bomb. In June 1953 a columnist for the *Washington Post* referred to Teller "directing H-weapon research" at an undisclosed location, and eleven months later a staff reporter for the paper called Teller "in charge of the H-bomb project."[53] (By no means was Laurence the only one to exaggerate Teller's contribution. In 1954 *Time* reporters James R. Shepley and Clay Blair Jr. wrote *The Hydrogen Bomb*, which elevated Teller while claiming that the H-bomb program had been undercut by liberals and communists at Los Alamos.)[54] When Teller died in 2003, the *Times* referred to him as "a fierce architect" of the hydrogen bomb in its obituary, in an apparent effort to finesse the distinction between his perceived leadership of the project and the reality of his fraught relationship with it.[55] But it was too late to correct the distorted picture of Teller's role as the father of the H-bomb that Laurence had pushed decades earlier.

11

Atomic Dialogue

The United States was enmeshed in a full-out atomic arms race with the Soviet Union, and in the view of American weapons scientists, that required the United States to conduct myriad atomic test explosions. With a view to keeping ahead of the Soviet competition, some of the atomic tests refined the designs of current fission weapons; others sought to advance the knowledge for future fission weapons; and still others focused on developing techniques and technology for the all-important hydrogen bomb. Officials overruled concerns about the environmental cost of radioactive fallout and plunged ahead.

Faced with the high demand for atomic testing, the US government, which had been testing only in the South Pacific, decided to resume nuclear explosions in the continental United States. Senior military officers had begun to push back against atomic testing in the South Pacific because of its cost and logistical headaches, so the federal government inaugurated a new nuclear testing ground in Nevada with Operation Ranger, a series of five tests of devices dropped from bombers over the site's Frenchman

Flat from January 27 through February 6, 1951. They were relatively small explosions that focused on weapons designs using smaller amounts of nuclear fuel.

Laurence was slow to cover the Ranger blasts. Admittedly the AEC did not make it easy for Laurence or any other journalist: the AEC did not announce the timing of the five explosions and made no arrangements for access for journalists to observe them. The *Times* ran articles by the Associated Press, as well as the *Times*'s Washington bureau and its West Coast reporter, who reported from Las Vegas. Only after the fifth and last Ranger test did Laurence file a story. In his front-page analysis, filed in New York, Laurence concluded that Ranger likely tested technology for atomic-tipped "artillery weapons and guided missiles." He based those conclusions on his own analysis and cited no sources.[1] Although the tests' low-yield explosions were used to develop low-yield tactical weapons, a key point missed by Laurence and other reporters was that another key aim of the tests was to check atomic designs that were to be incorporated into the subsequent Greenhouse series of four explosions at Enewetak Atoll in the South Pacific.

The AEC said little about Operation Greenhouse in advance, and no journalists were allowed to attend, but Rep. F. Edward Hébert was designated an official congressional observer for the Greenhouse tests.[2] Greenhouse commenced April 7 with a weapons-related test, which was followed on April 20 by another. Both escaped notice by the *Times*. But on May 8 the George test in the Greenhouse series achieved a landmark when it demonstrated that a fission explosion could trigger nuclear fusion by compressing a mixture of deuterium and tritium to the pressure needed for fusion in the moment before the fission blast swept it all away. George was not a fusion device; its fission yield was far more than the energy released by the small amount of hydrogen in it. But when the explosion—the largest atomic detonation until that date—worked correctly in the predawn Pacific darkness, the United States had demonstrated the basic principle needed to make fusion weapons work. Sixteen days later, in the final test of the Greenhouse series, another experiment demonstrated the principle of boosting—adding a small amount of fusion fuel to a conventional fission weapon to sharply increase the device's yield.

Despite the AEC's tight blanket over publicity, word about George leaked out when military members and civilians watching the dramatic blast wrote letters to their families, which soon found their way to hometown newspapers. Speculation was so intense that on May 25 the AEC and the Defense Department jointly announced that the Greenhouse tests had included experiments related to thermonuclear weapons research. The next day Laurence parsed the announcement, writing that it "definitely does not mean that an actual hydrogen bomb of any size has been exploded." Rather, Laurence wrote, scientists likely had conducted an experiment to verify that a fission bomb explosion could create the high temperature and pressure needed to create a fusion reaction.[3]

Provoked by Hébert's decision to publish his own eyewitness accounts of Greenhouse, journalists fought back against the prohibition on eyewitness press reporting. At a June 13, 1951, press conference, journalists nagged AEC chairman Gordon Dean for access to the tests. The commission did itself no favors with journalists with its handling of the next series of tests in Nevada, code-named Buster-Jangle, which incorporated exercises with several thousand troops. Even the troop commander suggested that reporters could observe the soldiers' activities, but the AEC refused.[4] The commission offered little information to the public other than reassurances that the public would be safe from radiation. *Time* magazine characterized the agency's attitude as "we didn't invite you . . . and don't expect any information from us." As a result, the newsweekly wrote, "reporters were thrown back on guesses, suspicions, and plain washroom rumors."[5]

The first Buster explosion, on October 22, came hours before the White House announced that the Soviet Union had conducted its third atomic test. Gladwin Hill provided the *Times*'s initial coverage of the first Buster test, called Able. Laurence weighed in a day later from Chicago, where he was covering a physics conference. Quoting comments by Dean from two weeks earlier, Laurence asserted that the Buster tests were focused on demonstrating weapons that were much smaller than existing weapons.[6]

Hill picked up the *Times*'s coverage of the remaining Buster tests, including a last-minute abort of the second test.[7] Laurence instead wrote stories on cosmic rays, the use of bread during a nuclear war, and the 1951 Nobel Prizes. He criticized the AEC's secrecy during a conference of the American Public Relations Association in Philadelphia in November 1951, about midway through the Buster-Jangle series. "I think I know something about the bomb and I can tell you that there is no valid reason for barring the press in Nevada," Laurence said. "I don't say that everybody should be let in on atomic tests, but trained writers should be on hand to keep the public informed."[8]

Bowing to the complaints about media access, staff members at the Atomic Energy Commission agreed to admit 375 journalists to observe an atomic test in Nevada during the next test series from April to June 1952, code-named Tumbler-Snapper.[9] The highlight was the Charlie test on April 22, in which a bomber dropped an atomic bomb over the Nevada desert. Simultaneously the military staged Desert Rock IV, maneuvers in the desert with more than 7,300 soldiers. The fact that the Charlie test was announced in advance enabled television stations to cover the event live, a first. Laurence wrote a front-page advance story on Charlie, running on the morning that the blast was scheduled. Laurence emphasized that it was to be more powerful than any other nuclear device exploded in the continental United States by that date. That contrasted with comments by Dean, who downplayed Charlie's power by emphasizing that the United States had exploded more powerful devices in the Pacific, and in fact would have exploded Charlie there too if it had been more powerful.[10]

On the morning of April 22, television cameras lined up to capture the blast. Afterward, Los Angeles station KTLA buttonholed Laurence to help explain what had happened. A KTLA reporter asked Laurence, "Did they split a little piece of the universe this morning?" Perched on the grass next to his interviewers, Laurence replied, "They did! They did!" But Laurence's explanation ran on at some length, leading the interviewer to bring the segment to a close. (After the interview, Laurence can be heard saying off camera, "Thank you! I hope I gave you something! I have to go back to write my story.")[11] In his story published the next morning, Laurence was no less enthusiastic about the test than he had been on television, noting that 2,100 troops safely moved through the detonation zone after the blast. Much of his article was devoted to a first-person account of the blast, which seems superfluous given the live television coverage.[12]

As Marcel LaFollette points out in her history of science on television, the Charlie event illustrates the change in science communication wrought by television, undercutting the primacy (indeed, the virtual monopoly) of newspapers in communicating facts and impressions about scientific news up to that time. This also undermined the impact of writers like

Laurence. "No matter how accurate, comprehensive, and insightful the prose, no matter how well-explained the physics, the impact of printed articles could now by eclipsed by the actuality of the visual," LaFollette observes.[13]

However, other reporters still regarded Laurence as a technical expert; the *Times* of London was one of the many newspapers that quoted his judgment of the size of Charlie's blast. One reporter recalled that when a group of journalists asked Laurence for his estimate, "the gnomelike little Laurence . . . squinted thoughtfully and estimated the bomb yield had been the equivalent of 15,000–20,000 tons of TNT. I don't know whether he had inside dope or was a good guesser, but that's what it turned out to be."[14] Laurence was in error yet again: the federal government today officially estimates Charlie at 31 kilotons, larger than the approximately 21-kiloton explosions that Laurence had already personally witnessed at Trinity, Nagasaki, and Operation Crossroads.[15]

Meanwhile, at the *Times*, editors and executives were beginning to worry that Laurence's political beliefs were tainting his news coverage about the atom bomb. For example, they were concerned over contradictory reporting by Laurence and Baldwin after the Truman administration announced on October 3, 1951, that it had detected evidence of a second atomic explosion by the Soviet Union. Laurence wrote a front-page article that asserted that the announcement was evidence that the Soviet Union's "progress in designing improved atomic weapons has been much slower than that of the United States." Laurence argued that the Soviet Union had twenty to thirty older nuclear weapons, comparable in destructive power to the weapon used at Nagasaki, and that the US nuclear stockpile was larger and more advanced than the Soviet stockpile. Consequently the Soviet Union could not mount a credible nuclear challenge to the United States, he claimed.[16] Simultaneously, Baldwin filed a story taking a contradictory view that the Soviet test was evidence of a robust nuclear weapons program. He wrote that the test suggested that the Soviet Union's stockpile comprised thirty to eighty nuclear bombs, each with a destructive power of about 20,000 tons of TNT, or similar in destructive power to the bomb used over Nagasaki.[17] (Although detailed information on the history of the Soviet Union's nuclear programs remains scanty today, one group of scholars estimates that the Soviet Union had twenty-five warheads in its stockpile in 1951 and fifty in 1952, and was making great progress, so Baldwin's view appears to have been more accurate than Laurence's.)[18]

Laurence and Baldwin had differed on atomic issues since Operation Crossroads in 1946, but their dueling estimates in the wake of the October 1951 announcement caused a conundrum: two of the paper's top reporters on nuclear-related issues were drawing diametrically different conclusions from the same event. The editors decided to run Laurence's article first, on October 4, accompanying coverage of the White House announcement, and to postpone Baldwin's article for one day. Baldwin complained to Catledge that "the question of atomic energy has gotten out of the field of the scientist and very definitely into the field of intelligence and the military"—areas where he, not Laurence, had expertise. "Bill tends to take a much more optimistic approach about Russian atomic progress than I have found the official sources that I speak to in Washington do."[19]

On October 5 Baldwin's article was buried deep within the paper on page 13.[20] Catledge wrote Baldwin to defend the way his story was handled. "We have got to work out some arrangement whereby you and Bill Laurence will not be at odds on the atomic question. We certainly do not want you to change your views and we do not want him to change his, but if you do have different views, The Times must find some way to present these views without confusion to its readers."[21]

Although *Times* editors were frustrated by the divergent views of their two prime atomic reporters, both remained on the beat; *Times* editors dispatched both Laurence and Baldwin to cover the 1952 atomic tests in Nevada. Laurence was told to focus on the news of the individual tests: "the explosion compared with the earlier tests, the progress in development of the atomic bomb as compared with earlier tests, any further research and prospects, etc." Baldwin, on the other hand, was assigned "interpretive" topics such as implications for new weapons and the military significance of the tests.[22]

In late 1952 the United States resumed atomic tests in the Pacific. With Operation Ivy, the Atomic Energy Commission held its first test of a full-scale thermonuclear device on November 1, 1952, code-named Mike. The blast was an enormous 10.4 megatons—by far the largest to that date, close to a thousand times more powerful than the Hiroshima bomb, and enough to obliterate the island of Elugelab in Enewetak Atoll. Designed by the brilliant physicist Richard L. Garwin, based on revolutionary new principles that had been proposed by Stanislaw Ulam and Edward Teller, Mike was not a bomb. Rather, it was a huge, ungainly device that was impractical as a weapon because it required liquid hydrogen—specifically, the isotope deuterium. Fully fueled, the device weighed as much as 65 tons, and at the time it was the largest cryogenic test device ever built.[23]

But it worked, and the AEC could not keep the lid on that fact. Mike quickly became, in the words of the columnist Stewart Alsop, "surely, the world's most open secret."[24] In response to public speculation and debate about the Mike test, on a Sunday about two weeks after the test, AEC officials acknowledged that the Ivy test series had "included experiments contributing to thermonuclear weapons research." That story received front-page play at the *Times*, but Laurence was nowhere to be seen: Jay Walz instead wrote the *Times*'s main story on the announcement, and the paper ran a United Press sidebar on eyewitness accounts of the Mike blast. To explain how Mike worked, the *Times* had to resort to using a United Press story with the headline "Hydrogen Bomb Gains Power through Fusion."[25] Only the next day did Laurence surface with an analysis story, buried deep inside the paper, describing Mike (in dubious scientific terms) as incorporating a "test tube" model of a hydrogen device rather than a workable device itself. A working device would have to await large-scale production of tritium at a plant then being built in South Carolina, he wrote.[26]

The Atomic Energy Commission resumed its atomic testing in Nevada in March 1953 with Operation Upshot-Knothole, a series of eleven explosions designed to support weapons development. The AEC was forthcoming with some of its plans for Upshot-Knothole, in part because the explosions were conducted in conjunction with military exercises involving 20,100 military members, which would have been impossible to keep secret. The first test in the series, code-named Annie, was staged for broadcast on live television, and Laurence reported on it in person. Days before Annie, Laurence wrote that the test series would "advance the development of new and improved atomic weapons and this to maintain this nation's vast lead over the Soviet Union."[27] Two days later, Laurence wrote another advance story, this one keyed to the live television broadcast. "Information gathered here today from reliable sources makes it evident that it will be a much more efficient weapon, using considerably less than three-quarters of the fissionable material of the model that exploded over Japan," Laurence wrote.[28] When the blast happened, Laurence wrote in a typically breathless first-person account for page 1, "It was the light of a hundred suns rolled into one, a light not of this world. We counted three and took off our goggles. There before us, fearful to behold, was the ball of fire, a giant, iridescent sphere, a new start in the act of being created, changing shape and color at breath-taking speed."[29] In all of this, Laurence made little reference to the highly dangerous radioactive fallout being released by the atmospheric test on the soldiers and the surrounding area.

Upshot-Knothole was the last atomic test series that Laurence covered in Nevada. As his writing about Annie suggests, the disparity was growing ever wider between Laurence's reputation as the leading journalistic authority on nuclear weapons and the reality of what he was producing for the readers of the *Times*. His articles focused on canned themes such as the superiority of US weapons over the Soviet Union's, the drama of the tests, and widely known information about the purpose of the tests. He wrote less and less on increasingly important matters such as the military strategy that was driving the development of certain types of weapons (and thus conducting certain types of tests) and the environmental and public health cost of the radioactivity being dispersed from the explosions in Nevada, the Pacific, and elsewhere. At the *Times*, others such as Baldwin were picking up those themes and thus were providing more important atomic journalism.

But if Laurence was devoting less effort to serving the readers of the *Times*, he was trying harder than ever to supplement his *Times* salary with money from outside projects. In 1952 Laurence latched onto the growing popularity of paperback reprints of hardback titles. Books on the atomic age were among them. In 1946 Pocket Books had issued a paperback edition of *The Atomic Age Opens*, which noted Laurence's coverage of the discovery of fission in 1939 and quoted his landmark 1939 article on atomic energy for the *Saturday Evening Post* but otherwise gave Laurence no role in the story. In 1948 Bantam Books had published a paperback edition of John Hersey's *Hiroshima*, which Knopf had published in hardcover in 1946 after the text had been first published in the *New Yorker*. By the time Laurence considered the idea, Hersey's book had shown enormous popularity as a paperback; by 1955 Bantam had printed five editions. Sensing a market opportunity, Laurence had his agent propose to Knopf the idea of packaging *Dawn over Zero* and *The Hell Bomb* together as one paperback book. But the paperback publishers Bantam, Penguin Books, and Popular Library already had rejected the idea of reprinting *The Hell Bomb* as a standalone paperback, and Knopf replied, "We haven't the slightest reason to believe that any. . . [publisher] would be the least bit interested in either of Bill Laurence's books, or both of them together."[30]

Moreover, Laurence continued to write articles for publications besides the *Times*, sharing insights that he did not share with *Times* readers. In April 1953 Laurence wrote in *Look* that the H-bomb's enormous power would require it to be detonated miles above the ground, in order to spread the bomb's effects as far as possible. "Heat would reach the ground first, vaporizing steel and wood and humanity directly below the center of the explosion. For as much as 35 miles in every direction, irresistible fires would burst forth, everything that was inflammable would roar into flame simultaneously. For at least another fifteen or twenty miles further out, damage would taper off, but exposed people would suffer lethal burns, and any highly inflammable material would go up in flame."[31] However, Laurence reassured readers that the Soviet Union would not have an H-bomb anytime soon, writing,

> We have very good reasons to believe that Russia is still way behind us on the fission bomb and that her present fission bomb models are way out of date. We also have good reason to believe that, in trying to catch up with us, the Russians built a very inefficient plant that has bogged down. They are probably just beginning to rec ognize their fatal mistakes, but it will take them some years to correct them. So much the better for the peace of the world. For by the time they get on the right track, we shall have gone a long way ahead of them once more.[32]

Laurence's prediction was proven wrong by the Soviet Union only four months later, on August 12, 1953, when it tested its first thermonuclear device. Laurence was entirely absent from the *Times*'s coverage of that landmark event.

While Laurence's work focusing on atomic issues drew the most attention, his reporting of nonatomic news still carried considerable heft, as illustrated by a September 1951 incident in which his writing moved financial markets. At a time when color television was not yet widely available, the physicist Ernest O. Lawrence, one of the leaders of the Manhattan Project, developed a cathode ray tube screen that could handle both color and black-and-white television transmissions. Laurence gave front-page coverage to the development, stating that the device could allow color sets to be "mass produced at a cost only slightly above that of present black and white receivers" and asserting that the tube would settle a major controversy at the time regarding rival technologies for color television.[33] When the stock price of Lawrence's company rose sharply, the industry newspaper *Variety* attributed the increase to Laurence's coverage, while noting sulkily that *Variety* had reported on the new tube before Laurence did.[34] For once, Laurence's technological optimism proved well founded. Eventually the tube influenced the design of Sony Corporation's Trinitron television, which was noted for its bright images.[35]

In March 1953 Laurence made another splash, with an article in *Look* titled "You May Live Forever." In the article Laurence predicted that skin cells from a scar could be used to clone (although Laurence did not use that term) a human body using techniques from embryology through which researchers had used tissue from salamander tails to produce other salamander organs. As was typical of Laurence, he was hyping a valid scientific development—one that would eventually mature into what is known today as stem cell cloning. However, Laurence did caution that the technology for doing so would be unavailable for as long as a hundred years, meaning that people would have to store their scar samples in freezers that long. He even offered a name for this new technology: "phoenixology." [36]

Reaction to the article was mixed. "Laurence's forecast of life everlasting was disturbing and a trifle dreamy," E. B. White responded in the *New Yorker*. "*You're* not going to live forever at all; somebody else just like you will come along to do the living for you—an exact replica or reasonable facsimile, born of your scar tissue. You yourself are going to be just as dead as you always knew you'd be."[37] The syndicated columnist Bruce Biossat called Laurence's vision "staggeringly wonderful and thoroughly frightening." Control of the stored tissue would be crucial, he argued.

"Suppose the label came off your jar. You might spend a new lifetime trying to find out who you were before." Or someone might create several copies of himself or herself. "One can conceive of half a dozen Einstein's [*sic*] delving into nature's mysteries and perhaps coming up with different notions," the columnist stated. "Clearly, Mr. Laurence has awakened us to vast possibilities." Biossat's column was printed as an editorial in some smaller newspapers.[38]

Seventh-day Adventists in particular were disturbed by Laurence's piece. "The hope of entering into eternal life by way of the scientific route is a false hope. Only religion can promise eternal life, and science cannot take the place of religion," an editorial in a Seventh-day Adventist publication stated. "Mr. Laurence assumes that science has made it possible for you to live forever. But the Scripture declares that eternal life is a gift of God that is received by faith."[39]

But Laurence also suffered a major disappointment in work outside the *Times*, when what should have been one of Laurence's biggest appearances in the 1950s turned out instead as a bust: an appearance on *The Ford 50th Anniversary Show*, a live, two-hour television program broadcast live by both CBS and NBC on June 15, 1953, sponsored by the Ford Motor Company to celebrate the corporation's half century by showcasing the previous fifty years of American history.

The producer of the show, the agent and theatrical producer Leland Hayward, sought to showcase the full range of American culture, particularly a remarkable duet by Mary Martin and Ethel Merman. Hayward envisioned the program also including intellectual content, such as a scripted dialogue between two experts about atomic power, which led him to seek help from the AEC's Dean in May 1953. Hayward wrote Dean that he wanted to address "the future in atomic power. . . . It must be the most important thing that's happened in this Century." Hayward asked Dean to suggest an expert to work on the program and wrote that he "would be willing to engage almost anyone you thought reliable and competent to do any research, or anything else on the subject—like Laurence of The Times."[40]

Dean couldn't think of anyone who was more like Laurence than Laurence, and Hayward and Laurence eventually negotiated a contract for Laurence to play the part of "writer on atomic stories."[41] Laurence was paid $2,500, less than the sums paid to Bing Crosby ($15,000),

Mary Martin or Ethel Merman ($10,000 each), or Edward R. Murrow ($7,500)—but more than what was paid to the singer Rudy Vallee ($1,100).[42]

Overall the show was a huge success, but Laurence's involvement with it was a disaster. He first wrote a "rough preliminary sketch" of an "Atomic Dialogue" between him and Murrow. During the four-page exchange, Murrow would have posed questions to Laurence such as, "Where does all this atomic energy come from?" In the scripted conversation Laurence would provide lengthy explanations, based on an elaborate metaphor for energy use in the universe that he called "the Bank of the Cosmos." In Laurence's explanation, various physical and chemical processes gradually deposited energy into this bank over the eons, and with the advent of atomic energy, humanity had started making minute withdrawals from the bank. "With atomic energy we are slowly learning the secret of the plants, how they make food for life out of water and carbonic acid gas by harnessing the sun," Laurence would have lectured Murrow and the television viewers. "When we learn that supreme secret, we shall be able to make in abundance food stuffs and all the other vital things that only the plants can make, out of water and a common gas from the atmosphere and a few cheap minerals from the soil. In this way we shall abolish hunger and misery from the earth, and with it the false red doctrine that feeds on hunger and misery."[43]

Although the metaphor was innovative, it—like so much of Laurence's writing—was overheated and scientifically incorrect, and it was ill suited for the audience of a variety show. A second draft of the atomic dialogue was a bit more polished but retained the overwrought bank metaphor.[44]

Hayward rejected Laurence's proposal. By the time the program aired, the atomic dialogue had vanished from the program, replaced by a dialogue between cohosts Murrow and Oscar Hammerstein about America's role in the postwar era. Laurence never appeared on camera. In the final script, Murrow told Hammerstein, "We were talking the other day to Bill Lawrence [*sic*] of the New York Times, and he was drawing us a picture of the fantastic possibilities of the peace-time uses of atomic energy—all of the curing of diseases that for long have been regarded as incurable. The matter of curing disease in plants, for example, which would create a revolution because it would mean that there would be no more hungry people in the world."

The show was a hit. It included musical and dance performances by Martin, Merman, Frank Sinatra, and Eddie Fisher, and it also celebrated the history of the United States. The show was hosted by Murrow and Hammerstein. It was a milestone in broadcasting history due to the eminence of the cast and the ambitiousness of the program. The *Times*'s own reviewer praised it but made no mention of Laurence's nonappearance.[45]

Laurence told the entertainment industry newspaper *Variety* that he had retained a speaking part in the show as late as the evening before the broadcast, when he rehearsed until 3:00 a.m. Laurence told *Variety* that he was told that his part had been cut because the program was twenty minutes too long. In truth, the broadcast actually ran short, with a lengthy curtain call added at the end to eat up the extra time; moreover, the addition of the Murrow-Hammerstein dialogue to the program strongly suggests that the producer simply wanted Laurence out of the show.

Variety reported that Laurence found the experience "personally embarrassing because of the advance publicity when in actuality he was a victim of that old show biz adage—winding up on the cutting room floor."[46] Laurence had a point; newspaper articles previewing the program had promised Laurence as providing a look at the prospects for atomic power. One newspaper stated that Laurence would appear with Oliver J. Dragon, a puppet that had appeared on the *Kukla, Fran and Ollie* children's television show.[47] After the broadcast, Laurence's agent, Alan C. Collins of the Curtis Brown literary agency, wrote Hayward's office to seek "a settlement of this troublesome business over the part that our client William L. Laurence of the Times has played in the making of the Ford film."[48]

It is unclear from the documentary record whether Laurence's agent succeeded in prying additional compensation from Hayward. But as American progress toward an H-bomb accelerated, Laurence soon would find his attention directed back at the South Pacific rather than a television stage.

12

The U-Bomb

The United States had made tremendous technical strides toward a hydrogen bomb, but the scientists and engineers still had not yet produced a practical, deployable weapon. An atomic test in the South Pacific on March 1, 1954, code-named Castle Bravo, was both a tremendous success and a catastrophic disaster for the nascent US weapons program. Bravo also triggered a different type of chain reaction—a cascade of events that would undermine Laurence's standing with his editors and contribute to the end of his career as a daily-news reporter.

Previously, the Ivy Mike test on November 1, 1952, had demonstrated the feasibility of a thermonuclear explosion, but the device it used was impractical as a weapon because it relied on supercooled liquid hydrogen stored in a huge refrigerated mechanism. A practical weapon that could be delivered by a bomber or a missile needed to be much smaller, lighter, and more storable. Castle Bravo's compact fusion device demonstrated a technology that fit the bill. Rather than using liquid hydrogen as its fuel, it used a solid compound called lithium-6 deuteride; because lithium-6 deuteride is solid at room temperature, it does not need a bulky refrigeration apparatus. A small amount of tritium was also included. A fission bomb near one end of the Castle Bravo device exploded, releasing radiation that was used to compress and heat the lithium-6 deuteride. Under those conditions, the small amount of initial tritium fused with the deuterium to create neutrons, which converted lithium-6 atoms to more tritium to fuse with the deuterium, repeating the cycle and ultimately releasing a large amount of fusion energy plus more neutrons. As was true with most technical details of the hydrogen bomb, the vital role of lithium-6 deuteride in Castle Bravo was a closely held secret.

The Castle Bravo device, code-named Shrimp, also incorporated another classified feature. The initial fission bomb and the lithium-6 deuteride fusion fuel were both placed within a casing of uranium-238, an isotope of uranium not used in fission bombs because uranium-238 is unable to sustain the runaway chain reaction that culminates in a nuclear explosion. In the case of Shrimp, the exploding lithium-6 deuteride would generate a rainstorm of extra neutrons that, traveling through the uranium-238 casing, would force those uranium atoms to fission even in the absence of a chain reaction. That would add significantly to the total energy of the explosion. But the fissioning of the uranium-238 also would generate much more, and more radioactive, fallout than a simple hydrogen bomb without a uranium casing would create.

The Castle Bravo explosion, at 6:45 a.m. local time on an artificial island in Bikini Atoll, was extraordinarily powerful, the equivalent of fifteen megatons of TNT. That was triple what weapons scientists had predicted, because the scientists had erroneously assumed that only lithium-6 atoms would contribute to the blast, when in fact more common lithium-7 (which represented as much as 60 percent of the lithium inside the Bravo device) also added to the explosion. The *Times* noted the Castle Bravo test in an unbylined, front-page story that reported the AEC's announcement of the start of the Castle test series, stating that one of the Castle tests might have double the power of the Ivy Mike thermonuclear explosion.[1] (This was incorrect. None of the tests in the Castle series doubled Ivy Mike's punch, although Bravo was approximately 50 percent larger.)

As the understated nature of its announcement implied, the AEC had intended for details of Castle Bravo to remain as secret as possible. But the explosion vaporized vast amounts of coral and other materials, generating radioactive fallout that was blown by shifting winds first over the Rongelap and Rongerik Atolls, which were inhabited—but which were not evacuated for three days, allowing residents to be thoroughly contaminated. Then a shift in winds drove the expanding fallout cloud over a Japanese fishing boat named *Daigo Fukuryū Maru* (Lucky Dragon No. 5). After returning to Japan, the crew of the Lucky Dragon were hospitalized from radiation sickness, and one crew member died. The AEC at first tried to play down the incident, and Laurence and the *Times* cooperated: on March 12 the *Times* ran an article reporting an AEC announcement that 264 people had been exposed to radiation in the shot. "All are reported well," the article quoted an AEC announcement as saying.[2]

But the Japanese, who once again had found themselves on the receiving end of American atomic military technology, did not keep quiet. Japanese scientists made and publicized measurements of the types and amounts of radioactive contamination from Castle Bravo, such as fallout that was detected on the boat and in fish. The data showed that the contaminants included radioactive substances that were much more likely to have been produced by a fission bomb than by a simple hydrogen bomb. Two substances—atoms of the isotopes uranium 237 and 238—were particularly puzzling, because neither isotope of uranium was publicly known to be used in fission or fusion bombs.

Under public pressure, on March 24 the AEC widened its estimates of the size of the geographic zone affected by fallout from Castle Bravo, and President Dwight D. Eisenhower admitted at a press conference in Washington that scientists had been "surprised and astonished" by the size of the Bravo blast.[3] Meanwhile, the AEC pressed on with more tests in the Castle series. On March 27 the United States tested another device at Bikini, code-named Castle Romeo. Like Castle Bravo, Castle Romeo used solid fuel rather than liquid hydrogen. But the lithium deuteride in Castle Romeo contained natural lithium, with only 7.5 percent lithium-6 and the rest lithium-7, rather than a processed version enriched in lithium-6 as had been used in Bravo. The Romeo device, positioned on a barge in Bikini Atoll, exploded with the equivalent of eleven megatons of TNT, far more than expected, making it even today the third most powerful nuclear blast ever set off by the United States. The dramatic photos of Castle Romeo are among the most iconic from the era of atmospheric atomic testing. Once again, weapons designers had underestimated how much the lithium-7 would add to the bomb's power. And once again, the *Times* covered the blast only in a short, unbylined article written from Washington with little detail or insight into what had happened in the South Pacific.[4]

The public was becoming concerned about the public health threat posed by radioactive fallout. On March 31 Eisenhower tried to defuse the mounting controversy about atomic testing by holding a press conference with AEC chairman Lewis Strauss at his side. The press conference started out well enough, with discussion centered on the Castle Bravo test and the issue of fallout more generally. But then Strauss derailed the carefully crafted message by revealing that the recent Castle Bravo and Castle Romeo tests had enabled the United States to design megaton-size bombs that could demolish any city in the world.

"How big a city?" a reporter asked.

"Any city!" Strauss replied.

"New York?" the reporter persisted.

"The metropolitan area, yes," Strauss answered.[5]

Much of the press coverage of the event centered on Strauss's claims that the United States had developed a new capability to wreak monstrous devastation anywhere around the globe. As one historian has put it, "in his own backhand way, Lewis Strauss had quieted the nation's concern over fallout by raising the spectre of nuclear holocaust."[6]

After the press conference, Eisenhower took exception to Strauss's answer to the question about how large a city the United States was able to destroy. Eisenhower privately told Strauss, "Lewis, I wouldn't have answered that one that way. I would have said: 'Wait for the movie.' But other than that I thought you handled it very well." Strauss worried that Eisenhower's comment signaled disapproval by the president, but Eisenhower's press secretary, James Hagerty, reassured Strauss otherwise.[7] Eisenhower's mention of "the movie" was a reference to a color movie of the Ivy Mike explosion from two years earlier, which the Eisenhower administration was getting ready to declassify and release, in order to show the public and foreign nations just how powerful the device had been. The administration planned to release the seventeen-minute film on April 7. Selected journalists got an early view of the film on March 31, under the condition that they agreed not to mention it publicly until April 7, an arrangement called a news embargo.[8] However, the administration's careful plans for stage-managing the release of the Ivy

Mike movie ran aground when the columnist Drew Pearson independently obtained information about the film and wrote a column about it, with publication set for two days before the movie was to be shown to journalists. Since the cat was out of the bag, editors at the *New York Times* decided to ignore the embargo and publish their coverage of the film on April 1, the same day as Pearson's column was published. That in turn provoked CBS and NBC to broadcast the film on the morning of April 1, and wire services to distribute their stories. Eisenhower administration officials then canceled the embargo altogether, freeing all news organizations to publish the story.[9] When Pearson came under criticism for his column, he responded that he had had no embargoed access to the film and that claims that he had broken the embargo therefore were "untrue and libelous."[10]

In the rushed release of the Ivy Mike film, the Eisenhower administration lost control of the narrative about the fallout problem, and the images of the enormous blast combined with the ongoing news coverage of the Castle Bravo and Castle Romeo tests—each of which was more powerful than the Ivy Mike blast shown in the movie—heightened public worry about the threat posed by fallout.

At this point, Laurence somehow caught wind of the fact that both the Castle Bravo and Castle Romeo shots had used solid lithium deuteride instead of ultra-cold liquid deuterium and tritium. It remains unclear who tipped Laurence off to this information, but on March 30 Laurence had traveled from New York to Washington to meet with two senior AEC officials: James Beckerley, the agency's director of classification, and Shelby Thompson, chief of the agency's public information service. Three days later, on April 2, Laurence published a story that disclosed that the Castle Bravo and Castle Romeo shots had tested "two new models of hydrogen bombs." His story reported that the shots had demonstrated that lithium-6 deuteride could replace tritium in hydrogen weapons. As was frequently the case in Laurence's reporting, the story disclosed no sources. Despite the story's newsworthiness, this story appeared on page 4 rather than the front page.[11] On the front page that day, the *Times* instead ran a story about claims by Strauss that atomic research held major potential for benefiting agriculture. Although newspapers subscribing to the *Times* syndication service ran Laurence's story, other papers appear to have ignored it.[12]

Laurence was not the first journalist to mention the role of lithium in hydrogen weapons. Using lithium deuteride had been discussed hypothetically in public since at least 1950, and speculation about lithium in Castle Bravo was particularly common. Three days before Laurence's story appeared, United Press distributed a story speculating that the United States was using lithium in its hydrogen bomb. UP's Joseph Myler repeated the speculation eleven days later.[13] The day before Laurence's story appeared, the *Christian Science Monitor* reported that "both the United States and the Soviet Union apparently are using lithium . . . as a major ingredient" of their hydrogen bombs.[14] But neither Myler nor the *Monitor* specifically mentioned lithium-6 deuteride, and they both framed their reporting as speculation rather than as confirmed facts.

Laurence was correct that the move to lithium-6 deuteride was an important development; the substance became the centerpiece of all American thermonuclear weapons from then on. Edward Teller had advanced the idea of using lithium deuteride as early as 1947, and Laurence had had frequent contacts with the Hungarian physicist. Perhaps Laurence had known about the potential of lithium deuteride but had felt muzzled by classification rules, even though he had frequently noted the logistical problems in relying on refrigerated liquid tritium.[15]

Unlike other journalists who had mentioned lithium, Laurence drew the ire of the AEC for his April 2 reporting. AEC officials saw his article as a major security breach. In their view, Laurence's article was not simply speculating about the use of lithium-6 deuteride; speculation (even if it was spot-on) was not barred by classification rules. Instead, in the AEC's view, Laurence was reporting as fact information that someone had leaked to him, and that leaker would have violated the law.

Memos from the FBI and the Justice Department and appointment books for senior officials at the Atomic Energy Commission document their scurrying about in the hours and days after Laurence's article was published, trying to figure out how to handle the suspected leak. Suspicion focused almost immediately on Beckerley and Thompson, because of their meeting with Laurence days before the story's publication. Only a month earlier, Beckerley had complained in a speech to the Atomic Industrial Forum, an industry group, that the federal government was being too strict in keeping its atomic secrets, arguing that the United States should "stop kidding ourselves about atomic secrets" and should recognize the expertise of Soviet weapons researchers.[16] After Laurence's article appeared, Beckerley was ordered to review it for security, and he submitted a memo concluding that the article was "merely conjecture"—a conclusion his superiors immediately rejected. Beckerley also said that he had told Laurence to submit his article for classification review before publication, but Laurence had not done so. Beckerley and Thompson denied that they had provided Laurence with any classified information, a claim that AEC officials did not accept. An FBI investigator reported that AEC commissioner Joseph Campbell and AEC investigator David S. Teeple "feel certain that Thompson and Beckerley disclosed restricted data to Laurence and had attempted to cover it up by saying the article was not classified."[17] Teeple in particular was a hard-liner on security. Strauss had hired Teeple, a former Manhattan Project investigator, to aid in the security investigation of J. Robert Oppenheimer, and no less an authority than the official history of the AEC described Teeple as "a man known around Washington for his excessive zeal in security matters."[18]

Laurence's story must have been particularly upsetting to Strauss, with whom Laurence had cultivated a friendship. When Columbia University bestowed an honorary degree on Strauss in 1954, Laurence and his wife sent a congratulatory telegram offering "our heartiest congratulations on your well deserved newest honor."[19] For Strauss's seventy-fourth birthday in 1955, Florence designed him a special desk lamp. "This will shine a steady light on the great man's equations," she wrote.[20] The Laurences also gave Strauss a reproduction of a letter by Einstein. Strauss returned the favor by having the letter reproduced in a brass plaque; he attached one plaque to the lamp given him by the Laurences and offered Florence a second copy. Florence wrote that she would have it attached to a wooden cigarette box.[21]

Nevertheless, Strauss was known for his desire to tighten security on atomic weapons issues, and the apparent leak to Laurence was a glaring example of the abuses he sought to eliminate. On the day of publication of Laurence's lithium article, Strauss's secretary first attempted to telephone Laurence just after noon, but the reporter was out of the office. When Laurence returned the call at 3:30 p.m., Strauss was not available. Laurence tried again at 5:30 p.m.; Strauss's secretary noted in a diary that Laurence was "very anxious to talk" but that "LLS does not want to talk."[22] Nonetheless, at some point they did connect. Strauss told Laurence not to publish "any similar articles." Laurence's unwise reply that he had heard about the AEC's objections to his article only made Strauss more concerned, because it suggested that an insider at the AEC was feeding information to Laurence.[23]

That same day, the AEC's General Advisory Committee—a high-level panel composed of scientists with full security clearances, chaired by the physicist Isidor I. Rabi—was wrapping up a three-day meeting in Washington. At the end of the day, the committee turned to Laurence's lithium article published that day. According to partially declassified minutes of the meeting, the committee members "deplored this both as a terrible leak of security information and as very damaging to morale in the Commission's laboratories." The committee had "considerable discussion" of Laurence's story.[24]

Meanwhile, Laurence kept plugging away at the lithium story, despite the AEC's concerns. On Saturday, April 3, he again wrote about lithium deuteride, reporting that it was the key to future fusion weapons. Because lithium deuteride was inexpensive and easily obtained, Laurence wrote, "any nation possessing ordinary fission bombs may now have hydrogen bombs."[25] That Saturday, Strauss's assistant spoke with Laurence and told him that Strauss would call him on Monday. But Strauss was still playing hard to get. On both Monday and Tuesday, Strauss's secretary reminded Strauss to call Laurence, but the staff noted that they had no record of any such call from the chairman.

On Monday, April 5—three days after publication of Laurence's article—AEC commissioners declared Laurence's article to have been "a serious disclosure of classified information" and a possible violation of the Atomic Energy Act. Kenneth Nichols, the AEC's general manager, told an FBI liaison that "the positiveness and timeliness" of Laurence's article suggested that someone had leaked classified information to him. But within hours, Nichols telephoned the FBI liaison at 6:10 p.m. to ask the bureau to delay action because Strauss wanted to talk with Laurence on the matter. Strauss believed that he "could possibly get Laurence to reveal his source for the article." FBI officials, perhaps peeved by the switch, replied that they would wait for written instructions from the commission.[26]

The next day, Nichols provided those instructions, writing FBI director J. Edgar Hoover that the commission was "seriously concerned" over Laurence's article, which Nichols described as "essentially correct." Nichols specifically objected to Laurence's statements that tritium was no longer required for a hydrogen bomb and that the deuterium could be chemically compounded with lithium-6 to create a solid compound that could be stored at room temperature. Nichols said that the agency would support criminal prosecution; the commission would be canceling contracts related to its use of supercooled tritium, so the abandonment of tritium would become obvious anyway, Nichols wrote.[27]

The next day, Laurence goosed the AEC's stress levels even higher with a story reporting that the use of lithium-6 deuteride would make practical a cobalt bomb that could generate enough radioactive fallout to kill all life on Earth. The story, which the *Times* inexplicably buried on page 4, touched on two security issues for the AEC: the use of lithium 6 deuteride and the prospect of a cobalt bomb. Laurence reported that the devices tested in Castle Bravo and Castle Romeo could have been converted to cobalt bombs by simply replacing their casings with containers made of cobalt. The development of lithium-6 was the key ingredient enabling huge thermonuclear weapons including cobalt weapons, he added, sounding a theme guaranteed to upset the AEC's classification officers.[28] Shelby wrote Strauss, Nichols, and the commissioners to express deep concern. "We believe it possible that the Laurence piece may set off speculation on the cobalt bomb by other writers in the United States and abroad. This could lead to a situation of hysteria," Thompson warned.[29] They were right to worry; Laurence's article did trigger global consternation. The *Times* of London, for example, paired an article reporting Laurence's analysis with another article about the illness of the Japanese fishermen who had been contaminated by radioactive fallout from Castle Bravo.

Strauss and Laurence finally met in Strauss's office at 10:45 a.m. on Friday, April 9.[30] Laurence denied relying on a leak and insisted that he had deduced the information in his story on his own.[31] Strauss was unconvinced; Nichols telephoned the FBI that afternoon and instructed them to proceed with their investigation.[32]

Weapons scientists continued to be concerned about Laurence's reporting. According to partially declassified minutes of a GAC meeting in late May 1954, the committee again discussed Laurence's reporting of lithium-6 deuteride. After Rabi asked if the AEC regarded Laurence's article as "serious," Henry DeWolf Smyth—a Princeton physicist and

AEC commissioner—replied that "the timing and manner of expression indicated a leak" and that similar reporting by *Time* magazine was being investigated.[33]

But nothing came of the security complaints about Laurence's reporting. Less than a week after the initial complaint by Nichols to the FBI about the April 2 lithium article, A. H. Belmont—the head of the FBI's Domestic Intelligence Division, who, among other things, years later would oversee the FBI investigation into the assassination of President John F. Kennedy—concluded that Laurence might have been able to deduce the conclusions in his article without relying on leaks. As an example, Belmont pointed to Strauss's own comments at Eisenhower's March 31 press conference that the Bikini tests showed that the United States "could substitute a material costing dollars a pound for a material costing thousands of dollars an ounce. A person familiar with the basic theory might deduce he was referring to lithium-six deuteride and tritium." There was no point in prosecuting Laurence, Belmont concluded. Laurence's defense that he deduced the information on his own "would be most difficult to refute, and it could well be true that he did deduce much, if not actually all of it from publications and other public information," Belmont wrote.[34] Thompson would announce his retirement from the commission's staff a month later.[35] The federal government would not declassify the fact that Castle Bravo had incorporated lithium deuteride until 1974.[36]

Even though Laurence's reporting had stirred up trouble because of his reporting of the use of lithium-6 deuteride, he failed to uncover another classified feature of the new bombs. Analysts such as the physicist Ralph Lapp had noticed with alarm that the fallout from Castle Bravo had been more radioactive than they would have expected. By October 1954 the British physicist Joseph Rotblat had used information on the fallout to deduce the structure of the Castle Bravo device that had produced it, but British officials convinced him not to publish the information, to avoid harming relations between the United States and Britain. Instead Rotblat tipped off American scientists, including Lapp.[37] In November 1954 Lapp began to examine the issue with an article in the *Bulletin of the Atomic Scientists* that discussed the likely impact of fallout on Americans after nuclear detonations.[38] In February 1955 Lapp again wrote about fallout, this time in both the *Bulletin of the Atomic Scientists* and the *New Republic*, giving the issue greater visibility.[39]

Lapp and Laurence were, at best, wary of one another, and Laurence wrote nothing about Lapp's analysis. But other journalists took note of Lapp's views and began to investigate Castle Bravo more closely. Finally, on March 6, 1955, Edwin Diamond, a reporter at Hearst's International News Service, reported that the Castle Bravo device had been encased in uranium-238, and he explained that this change greatly increased the blast's energy while also producing highly radioactive waste products. In other words, Diamond wrote, Castle Bravo had tested a hydrogen bomb that was intentionally engineered to spew deadly fallout into the atmosphere and potentially across the globe. Diamond wrote that the device used in Castle Bravo was called a natural uranium bomb, or a U-bomb for short.[40]

The AEC refused to comment on Diamond's report. A spokesman would say only, "We never comment on components of weapons."[41] Strauss, the AEC chairman, told a television program, "So far as I know, there is no such weapon."[42] But media attention snowballed. *Time* and *Newsweek* drew attention to the purported U-bomb.[43] So did broadcast journalists. Ten days after Diamond's story appeared, Garnett D. Horner of the *Washington Star* asked Eisenhower at a press conference about recent news reports on the " 'bargain basement' U-bomb." Providing no new details but also no denial, Eisenhower referred him to Strauss.[44] Laurence and the *Times* as a whole remained silent about it.

Finally the Sunday edition of the *Times* developed interest in the issue and commissioned an article by Lapp. Daniel Schwarz, the editor of the newspaper's Sunday edition, asked both Laurence and Baldwin for their views on Lapp's article. Baldwin agreed with Lapp's conclusions and argued for publishing it. Laurence flatly and incorrectly stated that Lapp was wrong: "Everything accurate in it is old stuff; everything that is not old is either completely wrong or misleading." He continued, "It is possible (and I am touching on classified material which neither Lapp nor Hanson Baldwin may know) that the U-238 was there . . . but it is wrong to jump from this meager fact to the conclusion that the U.238 [*sic*] was there as a nuclear explosive." Schwarz asked Laurence to consult more with scientists, and the next day Laurence reversed himself and recommended publication.[45] Then Laurence flip-flopped once again, opposing publication. The issue was made moot when the Atomic Energy Commission decided not to clear Lapp's article for publication.

In June, confirmation (of a sort) of the U-bomb came from an unlikely source: AEC commissioner Willard Libby, in a speech in Chicago on the issue of atomic fallout. He referred to an unspecified nuclear device that could generate "ten megatons of fission energy."[46] The *Times* initially all but ignored Libby's speech, running only one paragraph from an AP story about it, on page 6, which like other media coverage focused on Libby's assertion that current fallout levels posed no threat to human health.[47] But scientists read between the lines of Libby's unclassified speech, concluding that the device he was discussing had to be encased in uranium-238.[48] Scientists tried to get various journalists to take up the banner and run the story; nine days after Libby's speech, Anthony Leviero, a reporter in the *Times* Washington bureau, did. Libby's speech, Leviero wrote, meant that the United States had incorporated uranium into its hydrogen bomb designs, creating a hybrid weapon that would have more power than an ordinary atomic bomb but would produce far more radioactive fallout than a simple hydrogen bomb.[49]

Meanwhile, Lapp kept picking at the issue. In the June 1955 issue of the *Bulletin of the Atomic Scientists*, Lapp argued that the radioactivity levels in Castle Bravo fallout reported by the AEC were much greater than would be expected from an ordinary H-bomb; rather, the device's radiation was more consistent with an H-bomb that also incorporated a large fission explosion, in excess of five megatons. He further concluded that the radiation would be so strong that people in the path of its fallout would need to take shelter for months rather than a few days, as had been thought.[50]

Laurence finally sought to debunk the U-bomb in the pages of the *Times* on July 10. In a story headlined as an "analysis," Laurence criticized "reckless speculation" about the U-bomb. "Authorities of the highest competence have assured this writer that no such uranium bomb exists or could exist, as a fundamental law of nature makes it impossible. This law is that Uranium-238 cannot sustain a chain reaction," Laurence wrote, even though a chain reaction was not necessary when the H-bomb supplied the neutrons. He additionally dismissed the supposed smoking gun of the uranium-237 in the fallout, writing erroneously that the explosion could not have transmuted uranium-238 into uranium-237.[51]

Less than two weeks later, the physicist William C. Davidon, cochairman of the Federation of American Scientists, published a letter to the editor of the *Times* challenging Laurence's analysis. "It is not clear what is 'pure fantasy,'" Davidon wrote, because the reactions by which uranium-238 could be used to produce uranium-237 were well known to physicists. "The only significance I know of to the presence of U237 in the fall-out is that it indicates a substantial amount of U238 was present in the thermonuclear explosion."[52]

In November 1955 AEC commissioner Thomas Murray delivered a speech in which he discussed the fallout that had beset the Lucky Dragon. Laurence wrote that Murray implicitly confirmed that the bomb had contained uranium-238 through his emphasis on the extensive amount of radioactive fallout, which a hydrogen bomb without a U-238 casing would not have produced.[53] Other journalists paid close attention to Laurence's interpretation. AP wrote: "A part of Murray's speech was interpreted by William L. Laurence of the New York Times as making it 'clear that the so-called hydrogen bomb is actually a very large fission bomb.'"[54] But after Murray's speech, Laurence told Schwartz that he still denied the existence of the U-bomb.[55]

The controversy escalated to a new level of visibility when the liberal journal *The Nation* published a withering indictment of Laurence by the investigative journalist Gene Marine on January 28, 1956. Marine argued that Laurence had studiously ignored evidence of the device's existence out of deference to Strauss, with whom he was "on friendly terms." Marine also proposed that Laurence's coverage of Murray's speech, which had suggested that Murray had publicly disclosed classified information without permission, had amounted to a "spanking" of Murray on Strauss's behalf in retaliation for Murray's advocacy for more openness about American atomic weapons technology.[56]

Nonetheless, few journalists took up *The Nation*'s criticism of Laurence. One newspaper editor noted it in a column.[57] When another journalist questioned Strauss about it at a meeting of the Overseas Press Club and mistakenly referred to Leviero's story as Laurence's, Laurence—who was on the dais next to Strauss—"exploded wrathfully that it was not his story," the journalist wrote.[58] A longtime *Times* reader, a Mr. Irving Morrissett of West Lafayette, Indiana, wrote Sulzberger to express concern about the "dishonesty over a long period of time" described in the *Nation* piece. Sulzberger replied, "I am looking into the matter. My confidence continues to remain with Mr. Laurence."[59]

But management at the *Times* was much more concerned than Sulzberger admitted to Mr. Morrissett. Within days of the publication of the article in *The Nation*, Catledge asked Laurence to respond to the criticism, and Laurence produced a February 4 memo giving no ground. He wrote, "The so-called U-bomb is up to now a purely speculative weapon. No one in the know has ever as much as hinted that such a weapon exists, while Chairman Strauss insists that he has never heard of a U-bomb and does not know what it is." He called Lapp "a minor former physicist." Laurence admitted problems in his coverage of Murray's speech: "On second thought I regret having gone as far as I did, as nothing the Commissioner said confirms the so-called U-bomb with any degree of definiteness. I still strongly doubt whether the contraption really exists."[60]

That explanation did not satisfy Sulzberger. On February 11 he instructed Catledge, "It seems to me we ought to go a little further on this, because I am just not sure of Bill Laurence's story. Sometimes if he has not given birth to an idea he refuses to acknowledge that idea's existence."[61] Robert Garst, an assistant managing editor, turned to Baldwin, the paper's other expert on atomic weapons. On March 2 Garst wrote Baldwin, "The publisher is disturbed, as we are on the Third Floor, about that whole situation." Garst asked Baldwin what he knew about the subject.[62] Three days later, Baldwin replied with a two-page memo that contradicted Laurence. Baldwin wrote that he had known about the existence of the U-bomb since the fall of the previous year. He had heard about it from a "high-ranking military source" whose information was confirmed by the heads of two military services.[63]

On March 8 Garst summarized for Catledge what he had been able to put together and concluded that the U-bomb did exist. "For some unexplained reason Laurence is resisting the acceptance of this fact," he wrote. Garst seemed to look for a way to excuse Laurence's behavior, suggesting that confusion in terminology might have been part of the problem (and in fact multiple terms were in use for the advanced bomb, including U-bomb and fission-fusion-fission bomb). Garst added, "and it may be that Laurence is splitting hairs." *The Nation* had its facts straight, but "the personal attack on Laurence is unjustified," the editor wrote.[64]

Asked to recount what he knew about the issue, Leviero reconstructed the reporting of his story in a two-page memo with multiple attachments submitted to James B. Reston, head of the *Times*'s Washington bureau. Leviero criticized the November 18 article in which Laurence had reported that Murray had revealed technical details of the weapon. Leviero was unstinting in his criticism, writing that "Mr. Murray made clear nothing of the sort. Will somebody please read the Murray speech . . . and see if there is any detail of the nature and workings of this weapon. It isn't there. If it isn't there the presumption is that Mr. Laurence borrowed his elaborately detailed description of how the bomb works from the very scientists he had originally rejected. But now there is a recantation. Is this responsible journalism?"[65]

Today it is uncontested that U-238 is incorporated into some weapons in the US nuclear arsenal, although the design typically is called a "fission-fusion-fission" or "three-stage" weapon rather than a "U-bomb." The uranium is used to increase the total explosive power of the device.[66] What had gone so wrong with Laurence's reporting? Years later, Laurence told an interviewer that he did not report on the uranium-238 bomb because a source had lied to him on the matter, and Laurence trusted the source despite the abundant information supporting the existence of the U-bomb.[67] Given Laurence's close relationship with Strauss, and Strauss's evident interest in tamping down publicity about a US weapon designed to produce deadly fallout, the AEC chairman seems the most likely culprit for this part.

Laurence may have been susceptible to manipulation by Strauss because he was not reporting as hands-on as he once did, leaving him underinformed and without good sources. Laurence covered some of the atomic tests that the AEC staged in Nevada, but many other Nevada explosions were covered by other *Times* staffers. Gladwin Hill, the paper's Los Angeles bureau chief, covered multiple tests in Nevada rather than Laurence or other New York–based reporters because of his proximity.[68] And many of the US Pacific tests were covered by the *Times* from Tokyo or not at all.[69]

Baldwin offered a similar interpretation of the U-bomb controversy. In 1975 Baldwin said in an oral history interview that Laurence had fallen out of touch with current developments about nuclear weapons. "Bill Laurence sort of either rested on his oars or did not maintain the close contact with the people who were actually doing the developing [of nuclear weapons] that he had done while he with the Manhattan Project," Baldwin said. "Bill was in very good touch with many of the scientists like Oppenheimer, but, on the other hand, he was not in touch with the military people who were studying the military feasibilities of it and who were more or less sending to the Atomic Energy Commission the requirements for the weapons."[70] Eventually, Baldwin said, he asked retired air force general Elwood R. "Pete" Quesada to convince Laurence that not only was such a bomb possible, it had been developed. Quesada had unquestioned expertise, having been the commander of Joint Task Force 3, which had run the Greenhouse series of atomic tests that included the first thermonuclear test. "It was embarrassing for The Times because I used to refer to these things in my pieces and Bill just wouldn't believe it."[71]

Twelve years after the Castle Bravo test, the Georgetown University historian Carroll Quigley offered a more complex explanation for Laurence's intransigence. Quigley saw the development of the U-bomb and its ability to contaminate an enemy—even when delivered by relatively inaccurate ballistic missiles—as Strauss's response to the alarming news of the Soviet Union's H-bomb test in August 1953; in short, thanks to the U-bomb, even if the United States could not out-blast the Soviets, the United States still could thoroughly contaminate them. In Quigley's view, Strauss could not publicly admit the existence of the U-bomb or its rationale because that would have amounted to an admission that the Soviet Union's 1953 H-bomb success meant that the United States had fallen behind the Soviet Union in H-bomb technology. "This disadvantage had to be overcome as rapidly as possible, and the best way to do so was to shift from blast warfare to radioactivity warfare," Quigley wrote. Therefore, Quigley believed, Strauss convinced Laurence that the Bravo test was of a simple hydrogen bomb and that the U-bomb did not exist. (Of course, this alleged disadvantage did not really exist, as steady US progress in H-bomb technology demonstrated through the 1950s.) In a larger sense, Quigley saw Laurence as having been co-opted by the US Air Force and its allies, principally Teller and Strauss, to ensure the air force's continued role in atomic warfare. The army had successfully pushed for the development of smaller, tactical atomic weapons that it, not the air force, would use in battle, but in the era before long-range ballistic missiles had been deployed, the U-bomb would require air force bombers to be dropped. Because Robert Oppenheimer had opposed development of the H-bomb and had promoted the development of army battlefield nuclear weapons—both policies that ran against air force interests—the air force and its allies created political pressure against Oppenheimer, leading eventually to the revocation of his security clearance in 1954. Laurence's laudatory coverage of Teller as father of the hydrogen bomb helped reinforce the narrative that Oppenheimer had undercut that project, in Quigley's argument.[72]

Despite Laurence's behavior regarding the U-bomb, he continued to participate in the *Times*'s coverage of atomic weapons while the government was busily planning a series of tests of second-generation thermonuclear devices, scheduled for two years later in the Pacific, code-named Redwing. Eisenhower decided that a small pool of journalists would be invited to view one of the seventeen Redwing explosions, the first journalists allowed to witness a US atomic test in the Pacific since Operation Crossroads in 1946.[73]

Unsurprisingly, Laurence applied to be a member of the Redwing press pool, and in April the Capitol Press Gallery selected him and Marvin Miles of the *Los Angeles Times* to represent morning newspapers. Edwin A. Lahey of the *Chicago Daily News* and Thomas R. Phillips of the *St. Louis Post Dispatch* represented afternoon newspapers. Other journalists represented wire services, photographers, and broadcasters.[74] Strauss wrote Laurence to congratulate him on his selection and to compliment his performance on a televised discussion with youth about atomic testing. Laurence replied to thank him.[75]

Unlike with Operation Crossroads a decade earlier, the Redwing journalists were not treated to a lengthy cruise on a navy ship during which they could hobnob with one another. Rather, on May 2 Laurence and the other pool journalists flew on an air force plane directly from Hickam Air Force Base in Hawaii to the Kwajalein Atoll.[76] Once they arrived, the reporters were kept 50 miles away from the detonation site, in part to prevent them from seeing classified details of the Cherokee device. One reporter suggested in print that the device was a U-bomb design.[77]

Bad weather delayed Cherokee, giving the journalists a bonus: the unexpected opportunity to witness the Redwing Lacrosse atomic explosion, on May 4, 1956, while waiting for Cherokee. Laurence told his readers that Lacrosse tested "a relatively small device" designed to produce a much smaller blast than by-then typical hydrogen bombs. Laurence wrote that, based on his own personal expertise, he judged the blast to have been no more than 10,000 tons of TNT, or half the explosive power of the bomb dropped on Nagasaki. Laurence wrote that he drew this conclusion from the diameter of the fireball and the height and size of the mushroom cloud. Lacrosse proved, he wrote, that the United States had developed a new class of efficient weapons that required less nuclear fuel than earlier weapons.[78] At least one other journalist at the test site, Joseph Myler of the United Press, appears to have regurgitated Laurence's estimate of Lacrosse's strength, although within days Myler ratcheted up his estimate of the yield to more than 30,000 tons. Laurence never raised his published estimate.[79] As he had been at Crossroads, Laurence was quite wrong about the strength of Lacrosse's blast. The federal government now publicly estimates it as having been equivalent to 40,000 tons of TNT, or almost double the energy of the Trinity, Nagasaki, and Crossroads explosions that Laurence had witnessed.[80]

Laurence wrote even less in advance of the Cherokee test than the scanty amount he had written prior to the blasts at Operation Crossroads. On May 6 he predicted that the explosion would be "several megatons" (today Cherokee's actual yield is recorded as 3.8 megatons).[81] Laurence filed several short stories about weather conditions in the test zone.[82] By contrast, Miles wrote multiple advance stories on the weather, predicted the yield as being between three and five megatons, and filed a story reporting interviews with the crew of the bomber that would drop the Cherokee device.[83] Myler wrote a story that described Cherokee as testing "a giant H-bomb high in the air" that eventually could be delivered by long-range missiles.[84]

Cherokee originally had been scheduled for May 8 but was delayed to the early hours of May 21. "Each night, I left word with the watch to get me out at 4:15 a.m.," Laurence later told the *Times* employee newsletter, "and each morning for almost a fortnight—May 8 to May 21—a gob would shake me awake, only to say, 'It's 4:15 a.m., sir. You don't have to get up.'"[85]

During the preparations, the journalists enjoyed visits to much of the test infrastructure, according to Charter Heslep, a former journalist who had gone on to work for the Atomic Energy Commission and was on hand for the Redwing tests. "They saw more in one day than many of the people have seen who have been through every test out here," Heslep wrote in a letter to his family. "They were given aerial flights of Bikini and Eniwetok atolls. Then they were ferried to four different islands to view bunkers, instrumentation, camera stations, weather stations, etc. . . . The newsmen seems [*sic*] to be positively exhilarated. They are getting so much more than they expected."[86]

And with social events considered de rigeur, the journalists threw a party for Redwing officials on land on May 9. It featured swimming and dinner; Considine served as master of ceremonies. Laurence was declared to be president of the Eniwetok Proving Ground Local No. 1 of the Ancient and Honourable Society of H-Bomb Observers. Membership cards for the society were distributed to all present.[87] The card bore the title, "Ancient and Honourable Society of H-Bomb Observers" and certified that the holder "Shook, Rattled and Rolled." The cards were signed by

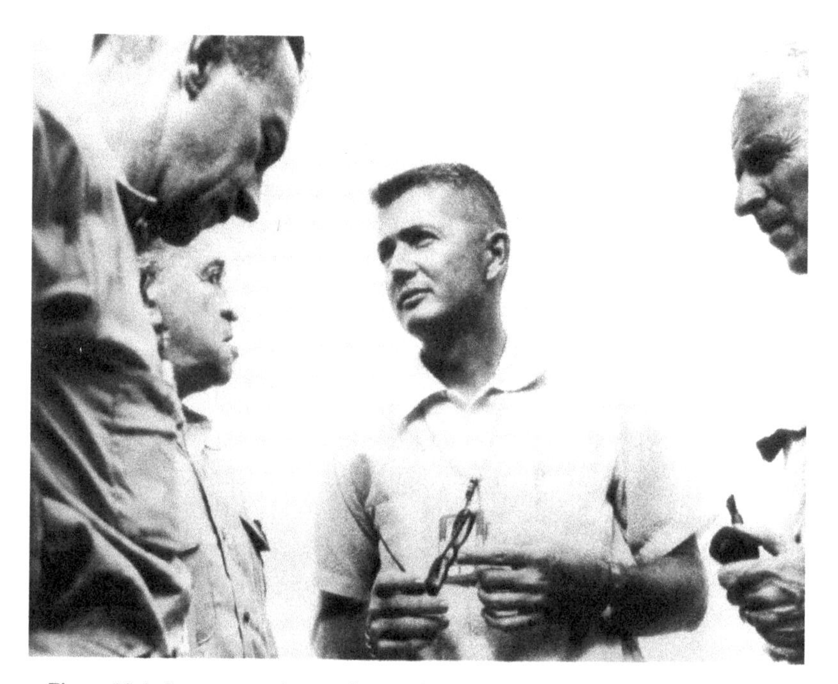

Figure 12.1. Laurence and two other pool reporters interviewed weapons scientist Gaelen Felt aboard the USS *Mount McKinley* on May 7, 1956, about plans for the test of a hydrogen bomb in the South Pacific. From left to right: Marvin Miles of the *Los Angeles Times*, Laurence, Felt, and Elton Fay of the Associated Press. (Associated Press)

"William L. Laurence, President, and Henry A. Renken, Chairman of the Horde by Order of Neptunus Rex."[88] Admiral Renken commanded the USS *Mount McKinley*, which played home to the visiting journalists.

On May 10 chances were estimated at 50–50 that the Cherokee test would occur the next day. "The newsmen got that tense look that comes when a story is about to break," Heslep wrote his family. But Cherokee was delayed yet again.[89] On May 15 the restless journalists staged a play, with one playing an admiral and four other journalists playing journalists asking the admiral for information, with little result. Laurence reprised his role from Operation Crossroads as King Judah, but Laurence's performance was "not so hot," Heslep wrote his wife.[90]

Unexpected news erupted when a bomber pilot was lost at sea after his B-57 went out of control. Laurence and the rest of the news delegation on hand wrote about the search. As Heslep observed in one of his letters home, "Ordinarily this story would get a paragraph. But because there is no other news and because they took part in it themselves, the boys are writing reams."[91]

While waiting for Cherokee, Laurence revisited Lacrosse, writing a story claiming that Lacrosse actually had been a test of a " 'pocket-size' hydrogen fusion bomb," sufficiently lightweight to be carried by a small bomber flying at high altitudes. The main evidence cited in the story was unexpected types of scientific instrumentation at the test location.[92] Today, the Lacrosse device is known to have been 34.5 inches in diameter, 100 inches long, and 8,386 pounds in weight. It was a prototype W39 H-bomb design, but it was tested with only its internal fission "primary" bomb and without its other nuclear fuel.[93] Thus Laurence was entirely incorrect: Lacrosse as tested was not a fusion bomb, it was over eight feet long instead of "pocket-size," and its measured explosive yield was four times larger than Laurence reported. Ultimately the W39 nuclear warhead was deployed in the B-58 Hustler supersonic bomber, the Snark intercontinental ground-launched cruise missile, and the Redstone short-range ballistic missile.[94]

The Cherokee shot finally took place in the predawn hours of May 21, the first US test of a hydrogen bomb dropped from a plane. Laurence wrote that his "personal observations of atomic explosions . . . strongly suggest that the explosion this morning was by far the most stupendous release of explosive energy on the earth so far."[95] According to Heslep,

Laurence first wrote that he estimated the shot at 25 megatons. "He is positively a fraud," Heslep wrote his spouse. "He took off into the wild blue yonder this morning and stated in his usual oracular way that the shot was 25 megatons—the biggest the A.E.C. had ever fired. The combination of his own fame and the prestige of the NY Times was such that even the Associated Press was pressuring its man, stable Elton Fay, to 'get with the story.' He told AP in NYC flatly that in his judgment, Laurence was flying a kite." According to Heslep, Rear Admiral B. Hall Hanlon, the commander of Combined Joint Task Force Seven, which ran Redwing, then announced to the press pool that Cherokee was smaller than any of its 1954 tests.[96] (That would have been a misstatement as well. The smallest 1954 test is now known to have been Castle Nectar on May 14, 1954, at 1.69 megatons.) Laurence tried to backtrack and insert a passage into his story adjusting his claim about the bomb's energy, but the printing presses at the *Times* had already started running, and night editors refused to make the change.[97] Today, the Cherokee blast is officially pegged at 3.8 megatons, much smaller than Castle Bravo or Castle Romeo, which Laurence had reported on, and far smaller than Laurence had claimed.[98]

Laurence's article described the explosion and its aftermath with great drama.

> For more than twenty seconds, the fire pillar and the fireball kept climbing at incredible speed. Then, as suddenly as it had appeared, the pillar of fire vanished behind the clouds. At the same time the fireball grew dimmer until it was no longer visible to the eye. Only less than half a minute had passed, but as far as human perception was concerned time had stopped flowing. One watched phenomena follow each other at incredible speed and yet felt as if they were taking place at slow motion. It was as if one were experiencing a nightmare with wide-open eyes in broad daylight.[99]

But Laurence used the Cherokee blast as a perch for again trying to debunk the U-bomb story. Laurence claimed that comments by an AEC official made clear that the Cherokee device was "the United States' first full-fledged thermonuclear weapon to be tested" and "a real fusion bomb." He added, "This lays to rest some fantastic speculations that the hydrogen bomb was not actually a fusion bomb but rather a so-called fission-fusion-fission device." In fact, Cherokee was precisely a fission-fusion-fission device.[100]

Laurence was not the only journalist guilty of grandiose prose about Cherokee. Miles, the *Los Angeles Times* correspondent, started his story, "Blinding brilliance burst the heavens wide open over Namu for an eternity at dawn today." The reporter estimated the blast at 15 megatons.[101] UP's Myler wrote, "The light that shattered the predawn darkness was greater than that of 500 suns. Its mammoth superhot fireball sent water and coral dust surging into a deadly radio-active cloud that spread 100 miles."[102] The International News Service's Bob Considine began his story, "America's first air-dropped hydrogen bomb was exploded with an enormous blast over the mid-pacific isle of Namu Sunday, proving the United States now has a 'packaged' thermonuclear thunderbolt it can hurl at any target on earth."[103]

All in all, the Cherokee test was less of a media event than earlier atomic tests reporters had been allowed to cover. One journalist, Philip W. Porter of the *Cleveland Plain Dealer*, complained that the media had produced "hardly a ripple of response" to the H-bomb explosion. "There was a certain sameness in the two bomb-drops, 10 years apart. Some of the same observers were out there—in particular, Bill Laurence, the wizened, learned science reporter of the New York Times and Bob Considine, the globe-trotting columnist of INS. They were able to report the contrast, and did ably."[104] Miles complimented Laurence's coverage. "Bill is 'over 30' and yet the only thing he couldn't do that I did was to climb a mast."[105]

Two days after Cherokee, the *Times* ran a follow-up AP story about it, rather than an article by Laurence, on its front page. The story sought to walk back Laurence's claim about the strength of the blast, with a *Times*-written insert in the AP story that said, "Revised reports stated yesterday that the test explosion was not the greatest ever recorded," and quoted Hanlon as saying that Cherokee's force was "substantially below" Castle Bravo's.[106]

Shortly after Cherokee, Laurence and the other journalists returned home; other nuclear tests were still scheduled for the rest of the Redwing series, but the reporters were not welcome to view those. That made Laurence's story a finale of sorts. Through the end of July, the *Times* would run eight more stories about Redwing tests, all with Tokyo datelines. In fact, since President Kennedy would end US open-air nuclear testing in 1962, the *Times* never again would have a reporter on-site for an atomic test by the United States in the Pacific.[107]

Officials at the Atomic Energy Commission were pleased with the press coverage of Redwing. They estimated that the journalists at the test generated 172,400 words of text, 56 still pictures, 3,500 feet of 16 mm film, and 180 minutes of radio broadcasting.[108] General Alfred Starbird, the AEC's director of military applications, told the commission that the media coverage had been "extremely worthwhile."[109] Hanlon, the commander of Joint Task Force Seven, said in a film report, "Judging from reports and comments it would appear that the press visit resulted in a substantial contribution to public relations for the military services and the Atomic Energy Commission."[110]

However, something in Laurence's Cherokee story upset someone in power. A note in Laurence's FBI file states that AEC officials reported to the bureau that the article contained unspecified "classified information." One possible culprit is the fact that Laurence's story stated that the Cherokee device incorporated lithium-6 deuteride, the same assertion that had upset the AEC in 1954. According to the FBI file, Justice Department officials concluded that there was not enough evidence to convict Laurence for any lapses in the Cherokee story.[111]

More consequential for Laurence was the fact that unhappiness with his Cherokee coverage extended to *Times* executives. "Laurence's story of the H-bomb explosion was overwritten to the extent of almost purple prose," Garst complained to a senior executive, Orvil Dryfoos, a few days after the Cherokee story was published. "This is not a new problem," Garst wrote, declaring that he had lost confidence in Laurence. "In many areas, Laurence has ceased to be a factual reporter and in some cases actually has become a propagandist in matters in which he is especially interested," Garst wrote. And then Garst dropped what, for a journalist, would be an atomic bomb of his own, recommending that they "warn all editors involved to be extremely cautious about Laurence's enthusiasm and to tone down his stories when other agencies such as the wire services indicate that Laurence has gone overboard."[112]

Three days later, Garst's critique was reinforced when Lapp wrote the *Times* to complain that Laurence's Cherokee story "contains many inaccuracies . . . and goes far afield from reporting or even interpretative reporting standards which the Times has so consistently maintained in its news coverage." Lapp specifically attacked Laurence's assertion that the "so-called H-bomb" does not include materials such as uranium (i.e., a

U-bomb). "Clearly, if the H-bomb were only a fusion weapon the radioactive hazard associated with its detonation would be slight. . . . It seems to me that the New York Times ought to print some explanation of its dual description of the superbomb i.e. Leviero's and Laurence's." Lapp appended a three-page, single-spaced list of comments on Laurence's Cherokee coverage. Lapp took exception to Laurence's claims of Cherokee's "most stupendous release of explosive energy" with a fireball of at least four miles and brightness exceeding five hundred suns, details about the device's design, and Laurence's discussion about the cost of the device.[113]

Lapp was even more critical in a private letter to Lapp's own editor. "I sent the editor of the NY Times a four page critique of what was wrong with Wild Bill Laurence's May 21st story from Eniwetok. It was a melange of weird self-contradictions. Pocket size H-bombs with a 0.5 megaton A-bomb trigger! Thousand-fold cheaper H-bombs with active material which was only tenfold cheaper! I felt compelled to counter Laurence's madness because he took such frontal issue with the superbomb for which I am credited with parentage extra-A.E.C."[114] Lapp would not let the matter drop and wrote another letter to the *Times* complaining about Laurence, arguing that "Laurence is not quite on the beam at present and I think you should be aware of the situation so that, if you wish, you may use caution in any Sunday story or editorial that Laurence may write."[115]

Two months after returning from the Pacific, Laurence finally issued a mea culpa, of sorts, regarding the U-bomb. On July 30, 1956, Laurence wrote an article, which was buried on page 19 of the Sunday edition, reporting on a speech by Strauss discussing the government's efforts to minimize fallout from hydrogen bombs. The key, Laurence concluded, was eliminating the uranium-238 incorporated in fission-fusion-fission bomb designs. In writing that, just over one year after he denied the existence of the U-bomb in a *Times* article, Laurence acknowledged his error—but only implicitly. Laurence also claimed that the U-bomb design had come about "by pure accident." The uranium-238 likely had been added to the weapon design, he wrote, not to generate radioactive fallout but to help contain the bomb as it exploded and to reflect escaping neutrons back into the exploding bomb's core where they could contribute to the explosive process.[116]

But even then, Laurence went too far for the AEC, with a claim in the story's seventeenth paragraph that likely made the eyes of most *Times* readers glaze over but nevertheless broke classification rules. The offending text was: "The elimination of this uranium 238 would mean that the fission products would become wholly limited to the relatively small amounts produced by the fission bomb trigger, a total of about fifty pounds." That specific weight cited in the paragraph was classified as secret, an AEC official wrote Hoover, requesting an investigation by the FBI.[117] But the AEC had to admit that hundreds of people had legitimate access to that information, and the information previously had been published in multiple classified documents, so it was unlikely that the FBI would be able to identify Laurence's source. The agency further conceded that "an individual with a nuclear physics background could, through educated speculation, arrive at Mr. Laurence's conclusions."[118] Those concessions led the assistant attorney general to conclude that the government would be unable to prove that Laurence had violated the Atomic Energy Act or espionage laws, and so the case was shelved.[119] Laurence had escaped retribution from the federal government for what he had published. But his work nonetheless had raised deep skepticism among his editors, setting the stage for deeper conflict just around the corner.

13

King Laurence

Throughout his career, Laurence had tried to serve multiple masters at the same time, to advance his own interest in promoting himself and adding to his income. By the mid-1950s his divided loyalties were beginning to cause serious friction between him and management of the *Times*.

Laurence's penchant for writing for competitors to the *Times*, or allowing publications to leverage his fame without sharing credit with the *Times*, was particularly problematic. He predicted in *Collier's* in June 1956 that researchers soon would develop cures for cancer, viral diseases, heart disease, and mental illness. The magazine promoted the article with full-page newspaper ads touting a "progress report by Pulitzer Prize winning science-writer, William L. Laurence," making no mention of his association with the *Times*.[1] *Times* editors may have missed that, but a month later they were astonished to see an advertisement in the *New York Journal-American*, a direct competitor to the *Times*, touting an article by Laurence scheduled to be published in an upcoming issue of the *American Weekly*, a Sunday magazine supplement for the *Journal-American* and other newspapers in the Scripps-Howard newspaper chain. The article, "Why There Can Not Be Another War," stated Laurence's view that a nuclear war would be so devastating that no nation could afford to wage one. Two days before the article was to run, the *Journal-American* promoted it in an advertisement on page 2: "Will there be a third world war? Noted science writer William Laurence, a Pulitzer Prize winner, believes that a nuclear war between the United States and Russia is out of the question. . . . Get Sunday's New York Journal American." The next day, the promotion ran on the newspaper's front page, saying, "Pessimists say a third world war is inevitable, but William Laurence—distinguished science writer and Pulitzer Prize winner—violently disagrees." Finally, on the day of publication itself, the paper ran on its front page a list of "Today's Exclusive Features," including "A Third World War? Out of the Question, Says Science Writer William Laurence."[2]

The article recounted Laurence's experiences observing atomic blasts in the South Pacific and at Nagasaki. It was larded with Laurence's usual dramatic flair: "Through high-density goggles I saw a super-sun rise over the vastness of the blue-black Pacific, with a dazzling burst of green-white light estimated to equal, for a brief instant, the light of 500 suns at high noon. Awestruck and unbelieving, I watched the enormous fireball expand in a matter of seconds to a diameter of about four miles." Although the phrase "mutual assured destruction" had not yet been coined, Laurence argued that perspective. "These earth-destroying weapons, now being constantly reduced in size and increased in power, thus make it certain that no nation, no matter how powerful, could dare risk a thermonuclear (hydrogen fusion) war, because such a war would mean certain suicide for the aggressor."[3]

Times executives were livid. "Turner [Catledge] gave Laurence unshirted hell; said this was a dirty game he's playing," one memorandum reported. Sulzberger wrote, "I find it very hard to understand him."[4] Lester Markel, the editor of the Sunday edition, declared, "Of all our episodes with him, this is the most startling."[5] Salt was rubbed in the wound four months later when *Reader's Digest* printed a condensed version of the article without any mention of Laurence's association with the *New York Times*.[6] The magazine ran newspaper ads promoting the article, stating, "Pulitzer Prize-winner William L. Laurence tells why, in the awesome light of an exploding H-bomb, one thing stands clear: thermonuclear war means certain suicide for the aggressor."[7]

A possible solution to this staffing challenge seemed to present itself to *Times* editors when the paper's science editor, Waldemar Kaempffert, died of a stroke on November 27, 1956, at age seventy-nine. Within a month, Laurence took the post of science editor. He no longer would handle daily science news coverage but instead would write editorials and an article for the paper's Sunday edition and be available for occasional news reporting. Senior editors proposed to raise his salary from $326.90 per week ($17,000 annually) to $18,000 per year, but he may have received a higher amount: a letter from Sulzberger to Laurence about the position has a blank where the salary was to be typed, and Laurence's personnel record omits his salary as of December 15, 1956.[8]

If *Times* editors believed that promoting Laurence was a clever way to solve their problems with him, they were rudely surprised. Less than one month after Laurence started as science editor, assistant Sunday editor John Desmond took him to task for writing editorial pieces that were too long. "We should cover 4 or 5 subjects there each week and try to vary the field dealing with medicine, atomic physics, chemistry, etc., so that we have material for the interests of many people."[9] The next month, Desmond complained that Laurence failed to meet his deadlines. "Let's get a new science editor," he advised Markel.[10] The month after, another editor complained that Laurence was missing the deadline for his Sunday stories.[11] Two months after that, the Sunday department turned down an article that Laurence had written about radioactivity, arguing that the piece was not balanced.[12]

In the midst of this criticism, senior editors were unsympathetic when Laurence in July announced his desire to cover four overseas scientific conferences in August and September 1957. "When the publisher appointed me Science Editor he stipulated that I should continue covering important scientific meetings," he wrote a *Times* editor, asking for "an early decision."[13] "Here we go again," Catledge complained to Orvil Dryfoos, the president of the *Times*, in July 1957. Catledge felt that Laurence was trying to usurp the editors' prerogative to choose which stories to cover. "I repeat, we would like to have him cover these things," Catledge wrote, but editors—not Laurence—should wield decision-making power.[14] *Times* editors negotiated him down to two conferences and also proposed that he operate on a per-diem rate, which would avoid him handing in receipts because he was notoriously lax in completing such paperwork. "It took more than four years to get an accounting from him on his atomic bomb trip to the Pacific," Garst wrote Catledge.[15] Editors suggested a budget of $800 for his trip, which Laurence negotiated up to $900.[16]

But Laurence's trip generated little news. While on the road, in mid-August Laurence encountered visa problems with the Soviet Union. Stuck in Paris, Laurence advised an assistant foreign editor in New York that a Moscow event was not going to be newsworthy and asked for permission to skip it. "Okay forget Moscow," the editor replied tersely.[17] In September Laurence visited Israel and filed three stories, which appear to be the sum of his output from the trip.[18] "King Laurence arrives Tel Aviv September 6 at 7[:]25," assistant foreign editor Nathaniel M. Gerstenzang warned the London bureau.[19] *Times* officials regarded the trip with "disappointment" and judged that Laurence had put in an "unproductive performance." One called it an "abortive trip." They discussed making Laurence pay charges above the $900 that he had been advanced, particularly for unauthorized side trips to Stockholm and Amsterdam.[20]

However, Laurence did manage to leverage the trip to Israel for his own benefit, when the next year he was invited to describe his impressions of Israeli science in a speech at the Theodor Herzl Institute in New York. He compared the growth of scientific research in Israel over the preceding decade to efforts to make a desert bloom. "Although scientists don't talk of miracles, what I saw there were literally miracles," Laurence said in his lecture, which the Herzl Institute later published in booklet form.[21] He praised Israeli scientific institutions, including the Weizmann Institute of Science, as being the equal of US institutions such as the Rockefeller Institute for Medical Research (today called Rockefeller University), and he pointed to research by Israeli scientists in areas such as water purification, photosynthesis, genetics, agriculture, and human biology.[22]

At about this time, Laurence's ego and insistence on being the leading journalistic expert on the Hiroshima bombing put him in conflict with the CBS television network when it developed a documentary on the bombing. The network claimed that the program "will disclose to televiewers, for the first time, how the specially formed 509th Composite Group, a B-29 unit, under Col. Tibbets' command, trained in complete secrecy at Wendover Field, Utah, in 1945."[23] Newspapers reprinted the handouts verbatim.[24] But Laurence already had reported the 509th's training in

Dawn over Zero, more than a decade earlier, and argued that the broadcast was based on his book.

The TV show, broadcast on March 9, 1958, made only brief mention of Laurence. After recalling the Enola Gay's takeoff from the Tinian airfield, Walter Cronkite intoned, "New York Times reporter William L. Laurence has asked copilot Capt. Robert Lewis to keep a flight log. It is in the form of a letter to his parents in Ridgefield Park, N.J." Cronkite then read excerpts from the log. Laurence objected to this, claiming that he owned the logbook, which Lewis had kept at his direction. But CBS rejected his complaints and rebroadcast the show on July 19, 1959.[25]

Tensions at the *Times* did not abate, and by September Desmond pleaded with Markel to cut back on Laurence in favor of Robert K. Plumb, a science and medical writer for the *Times* since 1947. "Let's get Plumb permanently in the science column and let Laurence write editorials," Desmond wrote.[26] Just before Christmas in 1957, Markel restated Laurence's deadline for copy for the Sunday paper; within days, Laurence had missed the deadline again. "Perhaps in view of the holiday there is some excuse. However, it would be good to jack him up," Desmond wrote.[27]

The executives' dissatisfaction with Laurence continued into 1958. In April Desmond complained, "Again it is Friday noon and we have no copy from Laurence except three or four notes."[28] By May 1958 tension with Desmond was severe. Desmond told Markel that Laurence was arguing "that he is the science editor and that I have no right to tell him what to do and how to do it."[29] The editors were getting fed up. Catledge denied Laurence permission to cover the second United Nations Conference on the Peaceful Uses of Atomic Energy, in Geneva in September 1958. Instead, the *Times* sent reporter John Finney.

Next, editors tried moving Laurence's deadline earlier in the week, to Wednesday. Desmond warned, "I repeat that if the copy is not ready we shall have to make other arrangements to take care of the science column."[30] One idea was to have a backup story on hand to run if Laurence missed his deadline, but that proved impractical because Laurence's column was expected to deal with current events, so editors would have to commission a new backup every week. "It is Laurence's job and he has to meet what the job requires," Desmond wrote Markel. He did not.[31]

At this time, a brand-new exciting science story showed up on the horizon: the exploration of space. Initially it was a small story: US scientists and military agencies tinkered unsuccessfully with small-scale projects, not in any particular rush to score accomplishments. When the United States and other nations in 1957 started planning the International Geophysical Year, a seventeen-month international scientific collaboration focusing on earth sciences, *Times* editors started planning a special section to cover it, and they agonized over involving Laurence. "I would rather pass him up but it would be quite a slap," Desmond wrote Markel. "If we get the copy early enough, we should be able to deal with it."[32] Laurence was assigned a story about American plans to orbit an artificial satellite during the year.

When the Soviet Union orbited Sputnik 1 on October 4, 1957, the space story suddenly morphed into a race between the superpowers for dominance in space. Many reacted with hysteria, but President Eisenhower was not among those, instead publicly downplaying Sputnik's significance and resisting calls for responding with a crash program. Laurence had the same view; less than a week after the Sputnik launch, he told a journalism group in Salt Lake City that responding to Sputnik with an accelerated US space program would be "disastrous." He claimed that a US satellite would have little military value aside from collecting weather data and would be far less important than the atomic bomb.[33] Two months later, Laurence was still downplaying the Soviet Union's accomplishments as "small potatoes," arguing that the Soviets had focused on minor accomplishments with propaganda value while the United States was focused on launching a satellite that would have true scientific merit. "We could have launched a satellite in 1954 or 1955," he was reported as saying by a Rhode Island newspaper, which referred to him as "Dr. Laurence."[34]

Laurence began to warm to space exploration by June 1960 once he found an atomic angle to the story: a proposal for developing an atomic-powered spacecraft that could bring astronauts to other planets by 1970. The proposal, and another for a robotic craft to photograph Mars, "called attention to the fact that interplanetary travel and exploration is approaching at a faster pace than is generally realized," Laurence wrote in a Sunday article.[35] On May 25, 1961, President John F. Kennedy made his historic speech before a joint session of Congress proposing that the nation land on the moon before the end of the decade, and by 1962 Laurence had evolved into an ardent fan of human space flight, telling a group at the Franklin Institute, "I can't understand, and I think Franklin wouldn't have understood, those who say, 'Is it worth $40 billion to go to the moon?' It *is* worth it, because the knowledge we shall get will be worth hundreds if not thousands of times more than the $40 billion we shall spend over the next ten years."[36] Even so, Laurence's stories for the paper's Sunday edition rarely touched on NASA's Mercury, Gemini, or Apollo programs involving astronauts.

One concept that did fascinate Laurence was the prospect of extraterrestrial life, undoubtedly because of its connection to his decades-old interest in life on Mars. In a 1961 panel discussion on the prospect of life in the cosmos, Laurence asked, "Assuming that there are superior civilizations, couldn't we learn from them spiritual values that would help us on this earth?"[37] Laurence also commented, "If we assume that there is intelligent life in the universe, it is also rational to assume that there are some civilizations ahead of ours, civilizations that have solved a lot of problems that would take us a thousand years—maybe the problem of cancer, maybe also the problem of perpetual youth, physical immortality. Now wouldn't that be one of the greatest threats, greater than the atom bomb or anything else? Nobody would die, nobody would want to get old?"[38] Laurence himself covered the event for the *Times*, without telling his readers that he himself was involved in it.[39]

Laurence even found a way to earn some money from the space program, by delivering a lecture about space on a phonograph record that was sold by a company called Spoken Arts, which specialized in such products. Laurence received a $250 advance against twenty cents per record sold. "If we sell 20,000 which is not an impossibility you receive four thousand dollars," Arthur Luce Klein, the president of Spoken Arts, had written Laurence in enticing him to make the deal.[40] In fact, the company sold 2,197 copies of the record, for total royalties to Laurence of $439.40. Sales trailed off after the 1969 moon landing, but one sale was recorded as late as 1987, ten years after Laurence's death.[41]

"We are now entering one of the greatest ages in the history of man," Laurence intoned in the recording. He provided a disquisition on the laws of motion and the differences of gravity on various planets. He provided a history of rocketry from Newton to the present. He talked about the importance of satellites for purposes such as navigation, communications, and weather. Laurence said that "eyes in the skies" could prevent a "super Pearl Harbor" with nuclear-tipped intercontinental missiles.[42] The industry magazine *Billboard* called the recording "an exceedingly interesting dissertation by Laurence."[43] An advertisement from a local bookstore in Ithaca, New York, said, "You feel you are standing on the forefront of the space age with the scientists."[44]

In the fall of 1959 Laurence published his third book, *Men and Atoms: The Discovery, the Uses, and the Future of Atomic Energy*, this time by Simon and Schuster. The book originally had five parts. The first, which accounted for half the book, recounted once again his history of the development of the atomic bomb, including extensive text reprinted from *Dawn over Zero*. Other sections covered the hydrogen bomb, more history of atomic physics, predictions on future atomic technology, and a primer on atomic science. Later printings included a sixth section, on the neutron bomb.

Simon and Schuster had no easier experience working with Laurence than Knopf had had on his previous two books. "He had absolutely no interesting recollections or anecdotes and was unable to write English at all," Michael Korda, an editor there, wrote decades later. "Much of his manuscript made hardly any sense. Working with such raw material was a pleasure in its way, particularly since Laurence didn't appear to mind or even notice that I was rewriting his book from stem to stern."[45]

Some reviewers panned *Men and Atoms*. Robert F. Christy, a theoretical physicist who had worked at Los Alamos during the war, complained that he couldn't force himself to read the entire book. "Parts of this wide-ranging story are fascinating, but the book lacks the underlying unity that might tie together what ends up as a strange collection of essays." Harry Truman, by that time in retirement, repudiated passages that stated that Stalin had been fully informed about the atom bomb and that Truman had discussed it with Churchill before Stalin. "That isn't true," Truman told an interviewer. On Trinity, "when [Laurence] says that Stalin knew, he did not. He knew nothing whatever about it until it happened." When Truman talked to Stalin about it, "he knew no more about it than the man in the moon." Truman also rejected Laurence's claim that Truman had allowed the Soviet Union to get an edge on the United States in hydrogen bombs. Laurence wrote that Truman did not approve the H-bomb project until 1950. Truman's formal order to proceed with the H-bomb was issued in January 1950, but Truman told the interviewer that he had ordered the AEC to proceed with the H-bomb "long before that." (One historian has found Truman's assertion on this point to be "arguable" and better described as "expansive thinking.")[46]

Still, given Laurence's public profile, the book made a big impression. Even Marilyn Monroe owned a copy. One reviewer called it an "utterly absorbing book." Bob Considine called it "a breath-taking book. Easy to read; impossible to forget." John Troan, a science journalist for the Scripps-Howard newspaper chain, said it is "well worth your reading." The *New Yorker*'s reviewer called it "a marvelous book, a combination of literary genres that offers delight and information."[47]

By this time, the public debate about nuclear weaponry had moved to the difficult question of banning aboveground atomic tests, which wreaked enormous environmental damage but which, atomic scientists argued, were crucial to developing new weapons designs and maintaining older models. Since 1958 the United States, United Kingdom, and Soviet Union had been adhering to a moratorium on aboveground nuclear tests while negotiations on a permanent ban proceeded; John F. Kennedy assumed the presidency in 1961 after committing to pursuing a comprehensive test ban. Given his public reputation as an expert on the atomic bomb, Laurence was much in demand for his views on the test ban. In August 1961 he appeared on the popular radio show *All America Wants to Know* to debate whether the United States should break the moratorium on atomic tests. Laurence joined Oregon senator Maurine Neuberger and Wisconsin representative Robert Kastenmeier to oppose the resumption of nuclear tests, while championing the tests were Lewis Strauss; Rep. Chet Holifield, chairman of the Joint Congressional Committee on Atomic Energy; and Connecticut senator Thomas Dodd. Laurence roundly criticized another controversial proposal: the neutron bomb, a device designed to emphasize the release of radiation that could kill soldiers over blasts that would destroy structures such as buildings.[48] Laurence also addressed that issue in the pages of the *Saturday Evening Post* in May 1962, arguing that worrying about the neutron bomb was pointless. "At the present stage of world nuclear progress, it is scientifically unlikely that anybody can perfect an N-bomb for nearly half a century."[49] (Laurence was incorrect in that prediction. The United States would begin developing a neutron bomb in 1964, and the necessary scientific principles had been understood since the early 1950s.) The magazine heavily promoted Laurence's article, without mentioning Laurence's connection to the *Times*. "Science writer William Laurence says it's impracticable, costly and may never be built," one ad pronounced. "Get all the details about the N-bomb in this week's Saturday Evening Post."[50]

Despite the *Times* editors' objections to his work for other media companies, the *Times* had no compunctions about capitalizing on Laurence's fame for its own purposes. For example, the *Times* featured him in advertisements for the New York Times News Service, through which the *Times* sold its stories to other newspapers around the world. "What'll they discover next?" trumpeted one such ad featuring Laurence's photo, used by the *Sunday Gazette-Mail* in Charleston, West Virginia, to inform its readers in 1961 that it would soon begin using *Times* stories. "You'll know more, you'll understand more about the startling new discoveries and inventions of our time when you read the reports of William L. Laurence, the New York Times noted science writer. Laurence is only one of the many experts in many fields whose writings will become a part of this newspaper in September when the New York Times News Service becomes a part of the Sunday Gazette-Mail."[51]

Even the Soviet Union thought Laurence had enough valuable information to assign KGB agent Oleg D. Kalugin in December 1962 to contact Laurence, with the aim of "obtaining any information in conversations with Laurence regarding American atomic tests," according to Laurence's FBI file. The file contains no further details, and through a representative Kalugin declined comment to this author.[52]

Laurence's public star status was even clearer on April 29, 1962, when Laurence and his wife attended a gala White House dinner honoring America's Nobel Prize winners. The guest list included other intellectuals and academics, including several from the *Times*. Laurence was seated at a table honoring Linus Pauling, two seats away from him. Others at that table included former MIT president and presidential science adviser James R. Killian, Alton Blakeslee of the Associated Press, and biochemist Edward Doisy. Florence Laurence was seated to the left of J. Robert Oppenheimer at a table honoring him; others at her table included Harvard president Nathan Pusey, Nobel laureate Ralph Bunche, and novelist William Styron.[53]

But scientists were growing less sanguine about Laurence. In 1962 a master's student at Syracuse University named Ann Jacob Stocker provided evidence that his influence was on the wane. Stocker surveyed a hundred scientists, randomly selected from a directory, about Laurence and other science journalists. With a response rate of 63 percent, the mail survey indicated that Laurence had a decidedly equivocal reputation among scientists. In response to a question asking respondents to list science writers who have done "an admirable job," Laurence was named nine times, more than anyone else. The runner-up was not one person but a group—the staff of *Scientific American* magazine, which drew six votes. Of the survey respondents, 60 percent reported having read articles by Laurence; he was best known for his reporting on atomic energy. But comments by respondents showed a deep streak of dissatisfaction with Laurence. One wrote, "I think he became a little overimpressed with his own importance and his writing fell off in recent years." Another wrote, "Laurence handled the A-bomb story better than his colleagues because he was the only reporter invited to the tests. He has been coasting on his Pulitzer prize ever since. His books are exaggerations. His columns now-a-days are rehashes of things written by others."[54]

14

Peace through Understanding

For any fan of science and technology, the New York World's Fair of 1964–65 was an irresistible destination, an ode to progress despite the 1960s' specter of atomic Armageddon. Sprawling across 2.6 square miles in Flushing Meadows, New York, the fair's exhibits offered an exciting vision of progress for the 51 million visitors who passed through its gates; computers, modernistic automobiles, and telecommunications were all featured prominently. NASA and the Pentagon exhibited actual rockets in the Space Park. Rides offered park visitors the chance to glimpse concepts of a future of "Peace through Understanding," the fair's theme.

For Laurence, the fair was all these things. But the fair also would be the cause of Laurence's professional ruin, as he once again tried to serve two masters—his professional responsibility as a journalist and his consuming desire to be close to those in power, in this case the powerful Robert Moses, the so-called master builder of New York who, among many other projects, headed the fair.

One of the most popular exhibits was sponsored by American Telephone and Telegraph, unveiling its new Picturephone, which foreshadowed today's conferencing systems like Zoom and Google Meet by allowing callers to see each other through live video paired with an audio connection. People in adjoining booths at the Picturephone exhibit loved talking with one another using the system. AT&T gathered user reactions to help fine-tune the service before rolling it out commercially.

None of the visitors to AT&T's booth could have been more enthralled with the Picturephone than Laurence. On a cool, rainy day in April 1964, two days before the fair officially opened, Laurence made telecommunications history by holding the first transcontinental videophone call, with Donald Shaffer, managing editor of the *Anaheim Bulletin* newspaper, who was stationed at a similar booth at Disneyland in Southern California. The two chatted amiably while reporters in New York and California watched, and then the two held a coast-to-coast press conference, with reporters at the Anaheim park peppering AT&T executives at the fair booth with questions about their plans for the picturephone business.[1]

But Laurence's Picturephone demonstration that day also underscored a major change in his career: he was not covering the Picturephone demonstration for the *Times* because months earlier he had retired from the newspaper, embarking on a new career promoting the fair. Moses, who had run a previous world's fair on the same site in 1939, hoped that Laurence's work for the 1964 event would attract international attention to New York City and provide an opportunity to construct needed infrastructure such as highways and parks.

When the 1964 fair's organizers hired him as the fair's president in 1960, Moses assembled a high-powered management team stocked with reliable friends and allies who could bring that vision for the fair into reality. Among the recruits was Laurence, even though he was still working at the *Times*. The two had been friends and occasional neighbors for years; the mercurial Moses appreciated Laurence's intellect, and Laurence reciprocated the affection. Ten years before the fair opened, Moses as head of the New York State Power Authority had appointed Laurence to a committee to advise the agency on the feasibility of using atomic power at a planned power plant on the US-Canadian border. The committee and Laurence recommended against using atomic power at the site, and the power authority followed its suggestion.[2]

Figure 14.1. Laurence in May 1960. At this date, he no longer was writing daily news stories about science and technology but instead was writing editorials and science-themed articles for the Sunday Week in Review section. (New York Public Library)

Moses was justifiably known as free-spending in his management of the fair, and Laurence—who continually complained that the *Times* was underpaying him—undoubtedly appreciated the substantial money that Moses steered his way. Laurence was always trying to move up socially: although he had lived in a tenement when he was a reporter at the *World*, by 1945 he and his wife lived in a five-room apartment on Seventy-Second Street near the East River. By the early 1960s they were living on the Upper East Side at One Gracie Square, a well-appointed co-op building where Moses was among their neighbors. Florence Laurence organized concerts in an adjacent park.

The fact that the *Times* had reported Laurence's involvement in the 1954 power study with no question or controversy probably encouraged Moses to involve Laurence again years later in the 1964 World's Fair.

Figure 14.2. Robert Moses in front of the Unisphere, the symbol of the 1964–65 New York World's Fair. Moses was the president of the fair in addition to exerting significant civic influence through other agencies and bodies. (Associated Press)

In the fair's planning phases, a key challenge was convincing the federal government to support the privately run event by providing high-quality exhibits at taxpayer expense that would help attract the planned 70 million fair visitors. To that end, Moses put Laurence in charge of a committee to write a proposal for federal support, working with Detlev W. Bronk, president of the Rockefeller Institute; John R. Dunning, Columbia University's dean of engineering; and Lloyd V. Berkner, president of Associated Universities, who were paid a total of $30,000 for the proposal.[3] Released in December 1960, their report called on the federal government to spend $30 million to build a Franklin National Center of Science and Education at the fair site, including $20 million for the building and $10 million for its exhibits. "A science exhibit officially sponsored by the leading nation of the free world must make it clear to all the world that in the great war of ideas [in which] we are now engaged, our greatest defensive weapons are not atomic and hydrogen bombs but the mind of man functioning in a climate of individual freedom," Laurence's committee wrote. It envisioned exhibits highlighting the scientific contributions of Americans, using television and motion pictures in the exhibit, and featuring hands-on exhibits that would allow visitors to recreate key scientific experiments.[4] The *Times*'s story about the proposal mentioned Laurence's participation, but the *Times* again took no action against Laurence.[5]

Although Moses was pleased with the report, it went nowhere with the administration of President Dwight D. Eisenhower.[6] Moses and his allies hoped that John F. Kennedy's administration instead would be more supportive. Laurence again crossed an ethical line, apparently without his editors' knowledge, by writing Jerome B. Wiesner, Kennedy's science adviser, to ask him to push Kennedy to select an amenable federal fair commissioner. At Laurence's request, Bronk also wrote Wiesner on the issue. An assistant to Wiesner wrote back to assure Bronk that the White House had registered Laurence's and Bronk's concerns and to promise that they were seeking "a man of national stature who can really provide leadership in the face of conflicting interests and pressures."[7] Moses, Laurence, and Bronk got the compliant commissioner they wanted when Kennedy appointed Norman K. Winston, a developer who was a close ally of Moses, for the position; Winston would support Moses's priorities for the fair, with few exceptions.[8]

Laurence again did Moses's bidding when he and his wife joined an overseas publicity trip organized by the fair. The trip, in January and February 1961, included stops in Jakarta, Indonesia; Singapore; Kuala Lumpur, Malaysia; Phnom Penh, Cambodia; Saigon, Vietnam; Hong Kong; and Manila, Philippines. In each location the Laurences and the rest of the delegation wined and dined local and national leaders to encourage them to participate in the fair. "Laurence, at the age of 73, is holding up beautifully," the head of the delegation, the former ambassador Myron M. Cowen, reported from Saigon. "He is a very sweet fellow. . . . The trip what with its rapid moves and heat and all day and evening schedules is providing [*sic*] to be a rather brutal one, and I have been urging Laurence to take it a little easy, but to no avail."[9] In Jakarta an interview on state radio between Laurence and Ben Grauer, a radio and television personality, went so well that it ran ten minutes beyond the planned fifteen. In

Phnom Penh Laurence was interviewed on peaceful uses of atomic power by Radio Phnom Penh, the radio service run by the US Information Service.[10] Officials at the *Times* knew about Laurence's participation in the Asia promotional trip; indeed, the paper ran a short unbylined article noting his involvement.[11]

After he returned from overseas, Laurence plunged even more deeply into fair planning. Laurence acted as an unpaid consultant to fair officials, offering suggestions for scientific and technical exhibits, especially (and most implausibly) an operating nuclear reactor on the fairgrounds.[12] He also touted the fair in a cheerleading interview with the entertainment industry trade newspaper *Variety*, saying that the fair would illustrate society's increasing technological abilities that "may add decades to the life of man" and technology for electronic diagnosis that will "accurately appraise the ailment, indicate its cure by scientific count, or project more accurately whether surgery or similar major correction should be necessary."[13] And in November 1961, Laurence discussed plans for the fair with a commission that was visiting New York to make recommendations to President Kennedy about federal participation.[14]

At some points, Moses seemed oblivious to the fact that Laurence did not actually work for him but rather for the *Times*. One such incident was in September 1962, when Moses was losing patience with what he judged as the Kennedy administration's reluctance to support the fair by providing an exciting fair exhibit about the nation's space program. Laurence—still employed by the *Times* in writing editorials—suggested that his newspaper could run a news story and an editorial that would pressure the administration to provide a space exhibit for the fair. Moses seized on the idea and had a letter hand-delivered to Laurence's home urging him to get the article and editorial done immediately. "There is no time left for a long campaign of education. The Times thus far has not been very effective on the subject of a U.S. Pavilion as a focal center at the Fair," Moses directed.[15] However, the *Times* does not appear to have run an editorial or news story on the subject. In August 1963 Moses asked one of his subordinates to have the *Times* assign Laurence to write a Sunday story about science at the fair.[16] It is unclear from the documentary record whether the request was ever passed on, but no such article by Laurence appeared.

Laurence's association with the fair also skewed his journalistic work. In the fall of 1963 Laurence got the opportunity to ask the first question at a press conference featuring Moses, but he lobbed a softball, asking Moses to describe how much science the New York fair would have. "Not as much as I hoped," Moses grumped, going on to complain about the federal government's decision not to build a science pavilion. The other two journalists at the event asked modestly more challenging questions on international participation at the fair and attendance projections, but then Laurence gave another soft one, asking how the fair would promote "peace through understanding." In Laurence's third question, he asked Moses to describe plans for the permanent science museum.[17]

Laurence's deep and conflicted involvement in the fair came to a head, with disastrous results for his journalism career, with a remarkable chain of events in late April 1963. Laurence first shared with Moses the text of an editorial he had written about the fair's science center, by then called the Hall of Science, that the *Times* would be running the next day; Laurence was able to slip the editorial into the paper because the editorial page director was out of town. His editorial argued, "It deserves the whole-hearted support of all who are interesting in making New York the cultural as well as the political, financial and industrial center of the free world." The editorial further declared that the hall "would serve as a nucleus for a permanent center of science and technology comparable in scope and function to the Lincoln Center for the Performing Arts."[18]

Anyone else would have been thrilled by such a ringing endorsement from the *Times*, but not the micromanaging Moses, who quibbled with Laurence's vision for the science center. "I have a hunch that Bill has become a little too excited about the Science and Invention features of the Hall of Science, with space only a very incidental matter," Moses complained the next morning in a memo sent to more than a dozen colleagues, including Laurence himself, arguing that the first priority had to be an exciting space exhibit for the World's Fair; developing a permanent science museum had to remain a second priority. "We can well lose the whole project by failure to grasp this distinction," Moses wrote. But he added, "We need Bill Laurence. He has been a great help from the beginning."[19]

The editorial about the fair ran on April 24, and the next day Laurence doubled down on ethical lapses by testifying before the city's Board of Estimate—an appropriating board for New York—to argue for funds for the Hall of Science. After the hearing, the board approved $3.5 million in construction funds.[20] The *Times* ran an article about the hearing, noting Laurence's testimony but not his connection to the fair.[21]

When John Oakes, the editorial page director of the *Times*, returned to the newsroom the next day, he was enraged by what had happened in his absence. Oakes confronted Laurence and followed up a few days later with a memo: "I am not too happy about your being Science consultant to the World's Fair, but since you have long ago accepted that position, I am not going to ask that you resign from it." But Oakes instructed Laurence not to take any pay for his work with the fair and to refrain from writing any more editorials about it.[22]

Laurence replied in his own memo that his testimony to the Board of Estimate had been purely "a public service" because he had received no payment for it. "My role as consultant is not related in any way to anything even remotely controversial," he asserted. (This could not have been farther from the truth. The Hall of Science was and would continue to be controversial. Some critics thought it should have been located in Manhattan rather than at the fairgrounds, a competing group had its own plan for operating the facility, and still others criticized the building's design.) Laurence held that his work for the fair was similar to "lecturing, the writing of books, contributions to periodicals, or radio and TV appearances." He noted that he was not paid a regular salary by the fair but rather a per-diem compensation when needed—as if this distinction reduced the ethical problem. He concluded rather remarkably, "In my 33 years of membership on the staff, there has never been the slightest doubt about my objectivity and my whole-hearted devotion to the highest standards of journalistic integrity."[23]

In the ensuing months *Times* executives moved to rein in Laurence's involvement with the fair. They had had no qualms about allowing Laurence to work on behalf of the army surgeon general and the Manhattan Project on programs that seemed to have universal public support, but the fair was another matter altogether. The fair was politically contentious because of its huge expenditures and use of public lands, so Laurence's freelance politicking endangered the newspaper's reputation. The issue came to climax in October 1963, when Charles F. Preusse, the fair corporation's attorney, informed Laurence that Oakes had written the fair corporation seeking to block it from employing Laurence. Laurence complained to Oakes, but

Oakes denied having written such a letter, and Oakes wrote to Preusse demanding an explanation. Preusse's reply did not explain but did describe Laurence's relationship with the fair in detail, informing Oakes that Laurence had been paid $5,000 in 1960 to lead the committee that proposed a US scientific exhibit at the fair. Since then, Preusse wrote, Laurence had collaborated with Moses to promote federal involvement in the fair but had received no additional compensation. When the fair corporation considered building a permanent Hall of Science that would continue after the fair's closing, "there were a number of discussions with Mr. Laurence," Preusse wrote. But Laurence's testimony before the Board of Estimate was on his own behalf and not on behalf of the fair corporation, Preusse asserted.

Preusse's letter also included at least one untruth: "Mr. Laurence has not been retained by the Fair Corporation as a consultant on the Hall of Science," he wrote.[24] Just months earlier the fair corporation and Laurence had signed an agreement under which Laurence was to be paid $200 per day, up to $10,000 a year, "as a consultant, on an independent contractor basis, in connection with development of the proposed Hall of Science at the Fair."[25] The arrangement was no secret, as the fair had issued a brochure trumpeting that Laurence had "been retained to create exhibits demonstrating in an exciting yet understandable manner, not only the laws of the natural sciences, but the application of these laws by industry to the needs of mankind."[26]

Even Laurence—so proud of his relationship with the fair—could not swallow Preusse's claim. He handed over a copy of his consulting contract to Oakes.[27] The next day, publisher Arthur Ochs Sulzberger Sr. notified *Times* executives in a brief memo that Laurence would be retiring at the end of 1963. As a sweetener, Sulzberger agreed to pay Laurence four extra months of salary.[28]

Laurence maintained that he was forced out because of his age—he was turning seventy-five in 1963. The author Gay Talese agreed; in his backstage memoir about the *Times*, *The Kingdom and the Power*, Talese stated that Laurence had to retire because of *Times* policies requiring automatic retirement at age seventy-five.[29] But Talese also wrote that the mandatory retirement age "was a rather vague thing at The Times, subject to all sorts of adjustments, depending on who was being retired."[30] In any case, a routine *Times* analysis showed that Laurence was making $1,587.45 monthly and would receive $735.45 monthly in retirement income.[31] The *Times* threw a retirement party for Laurence on December 16, 1963. Sulzberger worried about the expense, so the invitation list was limited. Nonetheless, dozens of colleagues from the paper attended, and Laurence later wrote Oakes to express his "heartfelt appreciation for your kindness and generosity in tendering me such a memorable and heart-warming send off."[32]

Because Laurence was such a journalistic fixture, his impending retirement was news, leading *Time* magazine to report, "When he steps out of Times harness next week, he will leave the paper's science department far stronger than he found it. Six Timesmen now patrol the beat, all of whom had the chance to watch a pro in action, and all of whom surely gained by the experience."[33] Moses practically salivated at the opportunity to have more of Laurence's time. On a clipping of the *Time* article, Moses scrawled to a subordinate at the fair, "Get Bill L. going."[34]

Seventeen days later, on January 1, 1964, Laurence retired. His last news story in the *Times* ran that day on page 8, describing the discovery of a hormone in the human placenta that could promote fetal growth.[35] Alongside it ran a news story about Laurence's retirement. "Mr. Laurence, known widely as 'Atomic Bill,' was one of those who founded the profession of science journalism," the unbylined article proclaimed.[36] It was thirty-four years to the day after he published his front-page story in the *New York World* that had helped pave his way to being hired by the *Times*.

Over the next few months, the *Times*'s news pages tracked Laurence's transition to life outside journalism, such as when science groups honored him, when he was named a scientific consultant to the March of Dimes, and when he was elected to the board of directors of the Thomas J. Deegan Company, headed by a former *New York Times* reporter, close ally of Moses, and executive vice president of the World's Fair.[37] More than three hundred people attended a retirement dinner for Laurence at the Statler-Hilton's Terrace Ballroom about three weeks after his retirement. "The spectacle of men like Detlev Bronk and Samuel Goudsmit and John Dunning vying with one another to fashion the most glowing, lyric oratory in praise of a mere journalist was a phenomenon to gape at," reported an article in the newsletter of the National Association of Science Writers.[38]

Despite retirement, Laurence kept writing, finally breaking into the paperback book market with the publication of *New Frontiers of Science* by

Bantam Books in 1964, an anthology of Laurence's *Times* stories. *New Frontiers of Science* addressed topics such as the exploration of space, life on other worlds, geophysics, astronomy, biology, and medicine—and of course atomic physics. Florence sent Moses a copy of *New Frontiers of Science* before it went on sale. Moses replied: "Thanks. I'm now drowned in science, or maybe just gasping."[39]

In early 1965 Laurence visited Alamogordo to research a twenty-year retrospective article on the Trinity test for *Esquire* magazine. With his usual dramatic flair, Laurence described the state of the Trinity Site: "My first impression on arriving at the site in broad daylight is one of disillusion. The country is desolate and silent. . . . A faded sign affixed to the fence evokes a ghostly echo of the tense enterprise that once took place within: 'Dangerous Area, Keep Out, By Order of the War Department.' . . . Beyond the padlocked gate, a small pocket in the center is all that remains of the blast's crater." The travelogue was interspersed with flashbacks to Laurence's experiences during the activities before and the day of Trinity, such as physicists assembling the bomb prior to its test. The article closed with Laurence picking up a piece of rock from the Trinity Site and bringing it with him to New York, where an expert confirmed that it still was highly radioactive. "After twenty years, Alamogordo still bears the mark," Laurence concluded.[40]

Laurence kept a connection to the *Times*, reviewing Lansing Lamont's and Stephane Groueff's books on Trinity for the *Times*, in neither case noting that he has been interviewed by the authors for their books. Laurence wrote that Groueff's book "does fill gaps in the story" while complaining that Groueff misstated some details.[41] In 1965 Laurence contributed an article to the *Times* Sunday magazine reporting interviews with key Manhattan Project scientists two decades after the atomic bombings. Laurence's article and other pieces by *Times* staffers were gathered into a book anthology named *Hiroshima Plus 20*.[42]

But Laurence's major professional activity after retiring from the *Times* was an expansion of his involvement with the World's Fair and its successor, the Hall of Science. Laurence worked from office space on the fairgrounds. He started in January 1964, days after retiring from the *Times* and just months before the fair opened on April 22, 1964. Laurence was Moses's general go-to man on all issues scientific and technical; for example, he escorted Russian officials on a visit to their fair, and he helped the Vatican develop an exhibit celebrating the pioneering genetics research of the Catholic monk Gregor Mendel. But Laurence's main focus was on developing plans for the Hall of Science, which would remain operating after the fair's conclusion. In May 1964 Laurence presented Moses with a one-page memo proposing $3.5 million for the hall. At first it would focus on a "General Introduction to Science," and later it would add one or more buildings covering a variety of scientific fields.[43] In January 1965 the fair corporation distributed to trustees a slick report proposing dramatic expansion of the Hall of Science after the fair closed. A key feature was text written by Laurence warning trustees that "the development of a Center of Science and Technology worthy of the world's greatest city will take many years and many millions of dollars." He proposed a Hall of Discovery for physical and life sciences and a Hall of Inventions devoted to applied sciences.[44] A *Times* article on the proposal reported Laurence as saying that he would be traveling to Washington to convince officials there to add the fair's Space Park to the museum.[45]

Laurence also maintained that one as-yet unfulfilled goal—operating a nuclear reactor at the fair site—no longer was out of reach, thanks to the cooperation of his friend, the nuclear chemist Glenn T. Seaborg, at that time the chairman of the Atomic Energy Commission. Dunning, Laurence, and others met with Seaborg in Washington, DC, on September 7 to ask for AEC support for the fair. Both Laurence's and Seaborg's separate written accounts of the meeting agree that Seaborg readily agreed on one request: permanently stationing the agency's highly popular Atomsville USA exhibit at the Hall of Science. But Laurence and Seaborg had very different recollections of how Seaborg replied to a second request, for a nuclear reactor. In his diary Seaborg recalled waving off that proposal, worrying about the cost—perhaps as much as $2 million—and the fact that it would require permission from the president and Congress. Other cities also might use the arrangement as precedent for asking for reactors of their own, Seaborg worried. As a result, Seaborg said that the Hall of Science was asked not to seek support from its congressional delegation for the idea.[46] By contrast, Laurence reported to his colleagues at the Hall of Science that Seaborg had agreed to request $1.5 million to build a TRIGA (Training, Research, Isotopes, General Atomics) reactor at the hall.[47] As often had been the case in his reporting at the *Times*, Laurence was overly optimistic in his predictions on the nuclear reactor. Although plans for the "atomarium," as the reactor exhibit was called, were announced in January 1969, the exhibit never materialized, derailed by financial and bureaucratic sand traps.[48]

The fair closed in October 1965, and the Hall of Science then had to financially stand on its own two feet. To help close a budget gap, one idea that arose was to cut Laurence's annual salary as science consultant by two-thirds, to $5,000.[49] In May 1966 Robert A. Harper—an assistant to Dunning—made what he intended as a friendly telephone call to Florence Laurence, to inquire whether her husband might be in the market for some additional freelance consulting work. Florence instead went through the roof, interpreting the call as a heavy-handed effort to push Laurence out of his museum job. Harper recalled, "Mrs. Laurence remarked about old people who are let out to pasture after they have done so much. This went on for several minutes without any response from me." Florence complained to Moses, who used a signature negotiating tactic and threatened to resign from the board if they did not keep Laurence.[50]

The Hall of Science was rededicated as a standalone institution separate from the fair in September 1966, and Walter Sullivan—who had succeeded Laurence as the paper's principal science reporter—gave the facility an equivocal review. "The city's Hall of Science dedicated at Flushing Meadow yesterday is not yet a true museum," Sullivan wrote. "Rather, it is a leftover from the recent World's Fair—and a hope for the future." The *New Yorker* was even more negative, noting the building's "grim" interior and leaky ceiling. By June 1967 the new mayor of New York, John V. Lindsay, proposed a new $10 million building to be built next to the existing Hall of Science building, including a nuclear reactor to be provided by the AEC. The reactor never was built. Contracting problems and New York City's financial crisis conspired to close the Hall in 1971 for a year and a half. Although Laurence had managed to wangle donations of the (non-reactor) AEC Atomsville USA exhibit and some of the equipment at Space Park—specifically, a Mercury-Atlas rocket and a Gemini-Titan rocket—the Hall of Science could not afford to maintain them. "Once gleaming majestic testaments to space travel, they became faded and neglected over the years," the *Times* reported in 2004. The two rockets were refurbished from 2001 to 2003 at a cost of $2 million. Despite having endured a series of roller coasters in funding and political crises, the Hall of Science today is regarded as one of the best science museums in the United

States, fulfilling a vision that Laurence enunciated in a 1964 oral history: "I hope to see that New York City gets one of the finest science museums of its kind in the world, which would give a comprehensive understanding of all science to all the people."[51]

Around 1967, Laurence convinced himself that he had been cheated out of much of his compensation for work on the Manhattan Project. The Manhattan Project had paid Laurence $25 per week for expenses. According to Laurence, James told him that the *Times* would also give his wife $25 in expenses each week and loan the family an additional $125 each week, so that she would receive Laurence's usual weekly income of $150 while he was away on the assignment, to prevent financial hardship for Florence. After William Laurence returned to regular duty at the newspaper after the war, he repaid the loan to the *Times*—a total of $2,125 for the seventeen weeks that he had been away. Essentially, he had covered the atomic bomb project at a steep discount on his usual salary. (Even so, as Beverly Ann Deepe Keever has noted, the arrangement was an ethical violation. The ethics code of the Society of Professional Journalists at the time forbade journalists from holding outside employment if it compromised their integrity.)[52]

Somehow in 1967 Laurence became aware of this error, from one sentence in *Now It Can Be Told*, a memoir published five years earlier by Leslie Groves, the general in charge of the Manhattan Project: "It seemed desirable for security reasons, as well as easier for the employer, to have Laurence continue on the payroll of the *New York Times*, but with his expenses to be covered by the MED [Manhattan Engineer District, the Manhattan Project]."[53]

From retirement, Laurence asked Sulzberger that the paper reimburse him for his repayment of the loan. "The Times has owed me this sum for all these years since September 1945, and should, in all fairness, reimburse me, together with the proper amount of legal interest," Laurence wrote Sulzberger in February 1967.[54] Laurence's demand caught the *Times* by surprise. Edwin James had died in 1951 and so could not help resolve the question. But Arthur Ochs Sulzberger, the son of Arthur Hays Sulzberger, was in charge at the paper, and he replied to Laurence that although his father was "extremely mindful of the great contributions that you made to The New York Times," Groves's book did not justify any additional payments to him. "As an officer of this company, it would be impossible for me on that basis and after such a length of time, to pay the money that you feel is due you."[55]

A year later *Times* executives were still debating the issue. Edward S. Greenbaum recommended that the *Times* resolve it by honoring Laurence at his eightieth birthday with "an appropriate cash present."[56] Catledge opposed that idea, doubting that Laurence had been short-changed.[57] Arthur Ochs Sulzberger also opposed it. "The war and Bill's contribution on the story of the atomic bomb were many years ago, and he had ample time prior to his retirement to collect any monies that were due him," Sulzberger wrote Greenbaum. Sulzberger also challenged the validity of Laurence's claim, writing, "I do not want to sound overly harsh, but I can assure you that Bill was not one to permit unpaid compensation to accrue."[58]

However, Greenbaum worried about the potential for bad publicity. "The Times is presented in the Groves' [*sic*] book in a much more favorable light than the facts warrant," he warned.[59] Sulzberger held firm, writing, "It would be a real mistake to make supplementary pension payments to a retired staff member years after his retirement."[60] Laurence never got the money.

As his concern about the missed compensation shows, Laurence worried about maintaining his finances in retirement. Living in New York was expensive, and writing gigs were few and not always well-paying. To reduce their living expenses, in 1968 the Laurence couple moved to Mallorca, Spain. One reporter wrote, "Bill Laurence (the Pulitzer-prized Times science whiz who was sole newshawk at the first U.S. atom bomb tests so many lives ago) and wife are retiring permanently to Majorca, where the livin's easy on $300 a month."[61] Many at the time shared this impression regarding life in Mallorca, an island in the Mediterranean Ocean east of the Spanish mainland, but living there was not truly so easy. In the 1950s the United States had extended credit to Spain in an effort to open up the economy of the totalitarian state. The easy credit helped trigger runaway inflation and currency devaluations, and when the Spanish government tried to control these, unemployment shot up, and the Spanish government pushed tourism on Mallorca.[62] Shortly after the Laurences relocated there, another journalist declared that "Mallorca is a perfect example of what can happen when tourism takes over an enchanted island and rearranges its character to extract the maximum profit from its natural attributes." Prices of products and meals far exceeded what would be charged on the mainland.[63]

The couple's move seemed ill-fated from the start. The newsletter of the Overseas Press Club reported, "Their beloved 16-year-old dachshund Asra died the day after their arrival, and Bill is recuperating at the Provincial Hospital in Palma after surgery, delaying the writing of his memoir."[64]

Even while in Mallorca, Laurence remained on the board of the Hall of Science back in New York. In June 1968 the board's executive committee voted to renew his consulting contract through the end of the year, and three months later he was reelected as secretary of the board. Finally, by July 1969, board members woke up to the fact of Laurence's relocation—"apparently permanently," as the meeting minutes described it—and selected a replacement secretary.

From Mallorca, the Laurences and Robert and Mary Moses maintained a correspondence that provides a window into the Laurences' circumstances. In November 1968 Florence wrote the Moses couple that medical problems had been preoccupying them: Bill had benign tumors removed from his left hand and a thigh, while Florence had an ulcer and a polyp that were being treated with a hormone. "Have no intention of being operated on—no matter what," she declared. To top matters off, Florence's brother had died, and her brother-in-law was gravely ill. Still, she looked for a silver lining. "Once we are finished with doctors we could live here rather well—the climate is wonderful and I hardly cough," she wrote. The medical issues had preoccupied them so they had not yet visited many of Mallorca's scenic spots. Their entertainment consisted of visiting the British Club after a doctor's appointment to read newspapers from Britain. Florence did the shopping, while cleaning was left to a domestic worker. "A wonderful woman—I understand her Spanish and she understands mine," Florence wrote. "Hope one day to get a good refrigerator. Then I could shop once a week—it will make things easier." She had to go to the market every day, waiting for a cab so that she could bring home her purchases. It was time consuming, tiring, and no fun.[65] For his part, Bill watched goings-on in New York from afar. For example, after the Hall of Science announced plans, which ultimately never came to pass, for adding a nuclear reactor in January 1969, Laurence complained to Moses that he had not received enough credit for the idea.[66]

By the end of their first year in Mallorca, the reality of the situation had settled in for Florence. She complained to the Moses couple in August 1969, "The place is full of tourists. . . . We couldn't get any milk for several days—the hotels just bought the supply." The Laurences' daily routine centered around listening to the news on the radio, reading international newspapers, and doing household chores such as marketing and watering plants. "This life leaves a great many voids, but with our finances it is the best we could do. The climate is good and our physical needs more than adequate. Hence a sense of contentment—and a feeling of peace," Florence wrote.[67]

Bill Laurence remained interested in the landmark atomic events of World War II. Despite Florence's dissatisfaction with Mallorca, he declined an offer to travel to New York to appear on a television program marking the twenty-fifth anniversary of the Hiroshima attack because, as he told Robert Moses in a letter, "it was to be only a short interview on the local CBS channel."[68] But for the twenty-fifth anniversary of the Trinity test at Alamogordo, he wrote a piece for Science Service.[69] After the logbook written by the copilot of the Hiroshima bombing mission fetched $37,000 at auction in November 1971, Laurence fired off a letter to the editor at the *Times* that rehashed his role in prodding Capt. Robert Lewis to keep the log. Laurence also revived his extraordinary position, originally floated in his 1958 dispute with CBS over its Hiroshima documentary, that he owned the intellectual rights to the material. He argued, "The log would never have been written had it not been for my suggestion. . . . All the material in it is published practically in full in my book, pages 220 and 221, and is fully copyrighted by me."[70] The *Times* declined to publish the letter, citing the competition for space from many letters.[71]

Laurence also continued to be a steadfast supporter of the use of atomic weapons at the end of the war, writing in 1970, "Even though atomic energy was used to destroy in the beginning, it ended the greatest war in history and saved thousands and thousands of lives, both American and Japanese. There is no question at all that if we had to invade Japan an estimated half million Americans and perhaps two or three million Japanese would have died."[72]

While in Mallorca, Laurence revived a theory of the origin of cancer that he had formulated decades earlier, that avidin, a protein found in egg whites, would block the growth of cancer tumors. In 1941 Laurence had published a paper in *Science* arguing that raw egg white could be used as a cancer-fighting substance. Laurence conducted no clinical or laboratory research on the question but wrote that he was expressing his hypotheses "in the hope that they may prove useful in crystallizing an idea" for researchers.[73] In a 1941 editorial, the *Journal of the American Medical Association* had noted Laurence's hypothesis.[74] A few experiments were conducted—one with financing from a friend of Laurence's—but the results were inconclusive. In Mallorca, with time on his hands, Laurence wrote another paper on the subject. Titled "Evidence in Support of Hypothesis Explaining Observed Spontaneous Recessions in Human Malignancy," the three-page, single-spaced document provided no new experimental data or theory. Most citations in the paper dated to the 1950s or earlier.[75] Laurence mailed a copy of the paper to his friend Detlev Bronk, president of Rockefeller University, who forwarded the paper to *Science*, which rejected it. Laurence was insulted.[76] To placate him, Bronk solicited a second opinion on the manuscript from William Trager, a biologist at the Rockefeller Institute, whose work Laurence had cited in the paper. Trager wrote that "it does not present anything new either by way of experiment or hypothesis."[77]

Bronk worried that Laurence's insistence on publishing his paper would backfire. "I hope you will not be discouraged," he wrote Laurence. "Your distinction as an interpreter of science through the press is so great, and indeed classic, that you should not risk this great distinction by publishing anything that is not well though[t] of by other competent workers in the field."[78] Laurence took Trager's rejection poorly, asserting that Trager "had read my article very superficially, as all his comments are totally beside the point." He asked that Bronk use his position as a member of the National Academy of Sciences to help him publish the article in the *Proceedings of the National Academy of Sciences*. "Any credit that may come to me will, of course, also be shared by you for your important role as a Socratic midwife," Laurence implored.[79]

When Bronk declined to act further on Laurence's behalf, Laurence turned to his new home of Spain and in 1975 published the paper in the Spanish-language journal *Archivos de la Facultad de Medicina de Madrid*.[80] Florence Laurence wrote Robert and Mary Moses that "we are both happy that Bill's approach to a new treatment of cancer was published. . . . It may help many people." Bronk, who received news of the publication from Moses, diplomatically wrote the Laurence couple that he was "delighted" at the publication. "It is good that this important contribution is now available to the medical profession."[81] But there is no evidence that Laurence's 1975 paper has ever been cited by another researcher in the scholarly literature.

Bill's health deteriorated. In January 1973 Florence reported to the Moses couple that Bill had been hospitalized for nineteen days and afterward was bedridden due to a blood clot in a vein in his leg. "We are both glad to be home—although he hates staying in bed," she wrote. "We miss our friends—and are pretty lonely—a bit of honest laughter is much needed—a *rare* thing here."[82] In May 1973 Moses, who had just written an article for the magazine *Travel and Leisure* on reading literature as a surrogate for traveling, suggested that Laurence write an article for the publication about Mallorca. "Pays well," Moses advised him.[83]

In early 1974 Florence wrote the Moses couple that Bill had been hospitalized for a "tired heart." She added, "P.S. Wish we were home."[84] A few months later, she updated the couple: "We arrived here six years ago today and I still feel like a stranger. Bill is more reconciled." High inflation was killing local businesses and led the Laurences to cut back on their spending. "We simply don't buy. Go to the same restaurants but don't order as usual so they don't make any more money on us. Have a cleaning woman twice a week for four hours instead of five times and go to self service laundry. They do the pressing for me. I don't know how to do it."[85] Seven years after moving to Mallorca, their American friends had not visited. "We had hoped to come home, but Bill's first operation precluded his returning and after that it was one thing after another," she wrote. "Nostalgia is an illness no one can cure."[86]

Laurence died at 1:30 a.m. on March 19, 1977, at the Clinica Mare Nostrum in Palma de Mallorca. The local death certificate listed "cardiac insufficiency" as the cause of death.[87] From New York, executive editor Abe Rosenthal tried but failed to reach Florence by phone, instead writing her a short note of condolence. "I just want to say for Ann [Rosenthal's spouse] and myself and for everybody who knew Bill that we all regarded him not only as a great figure in American journalism but as a good, kind and helpful friend."[88] Florence was heartbroken without her soul mate. "My life is grim and lonely," she later would complain to Robert and Mary Moses in September 1979.[89]

Laurence's lonely death during his self-imposed exile on Mallorca illustrated the futility of his decades-long efforts to cultivate the powerful in science, the military, and the government: in the end, all abandoned him. And in his desire to report what the powerful wanted reported, he in many ways ignored the needs of the *Times* and its readers, as well as the public at large.

In that sense, Laurence's legacy in modern science journalism, and his mothlike attraction to the light emitted by the powerful in society, is a continuing distortion of what science journalism could be all about. With the encouragement and approval of leaders in science and government, Laurence focused on the razzle-dazzle of science, with his unending efforts to characterize whatever findings he was reporting as the biggest, best, most significant, world-changing development in science or medicine. Meanwhile, he offered no scrutiny of the powerful in society whom he had befriended. That emphasis continues today: science writers still focus on new findings—sometimes hyped as "discoveries" by a particularly excited or inexperienced writer. Even when they are not writing about breaking news, these writers tend to focus on the newest results of the scientific process: Who found what? What do other scientists have to say about it? What scientific questions remain unanswered? Simultaneously, the science journalists largely ignore critical scrutiny of powerful people and organizations that make their profession possible, such as the operation of the machinery of science, in academia or industry, and its relationship with government. Certainly, exceptions exist, such as exemplary reporting on public-health issues during the coronavirus pandemic. But in general science writers give short shrift to vital public issues like the spending priorities of governments, philanthropies, interest groups, and universities, and they unquestioningly accept the scientific establishment's position that anything that is in its interest—such as higher spending on science—is by definition in service of the public interest as well. Only rarely do science writers examine fraud, plagiarism, and other misconduct in research. In short, today's science writers largely fail to follow the basic journalistic duty of watching how government and society work and then telling the public about it so they can make changes if they desire.

Imagine, for a moment, if political reporters in Washington wrote only about the text of the laws that passed Congress and ignored the operation of the political process. This is how science writers see their work. We have Laurence to thank for the example that he has set for the generations of journalists who followed him.

Laurence pioneered this approach to science communication, with his unending efforts to characterize whatever findings he was reporting as the biggest, best, most significant, world-changing development in science or medicine. The fact that Laurence referred to himself as a "science writer" rather than a "science journalist" is revealing: he wanted to tell people what scientists had accomplished, but he rarely showed much interest in bringing journalistic scrutiny to how well they were doing it, whether they were benefiting society, or whether their relationship with the government served the public interest.

A major part of the problem is that science writers tend to be fans of science, as Laurence was. "Not only do science writers like science, they want their readers to like it too, or at least to understand how important science is," the investigative science journalist John Crewdson wrote in a withering critique of science journalists in 1993. Crewdson could just as well have been writing about Laurence and his compatriots.[90]

Laurence naively saw science as an unalloyed force for good. "I believe that science not only will serve humanity, but it will save humanity," he wrote in 1959.[91] The risks posed by scientific advances worried him not a bit. "We are living in a new heroic age, greater than any of the past," Laurence wrote after his retirement from the *Times*. "The age-old struggle of Man against the Gods is raging full force in a thousand laboratories throughout the land, and in other lands as well. Instead of one Olympus there is a range of much higher mountains on which the battle is going on, every laboratory a super-Olympus, on which the most formidable citadels of nature are being stormed by mortals, their principal weapons the naked intellect sparked by the vision of a New Revelation."[92] In 1962, as the United States was getting ready to send a probe to the moon and John Glenn into Earth orbit, Laurence encouraged an audience at the Franklin Institute in Philadelphia to be optimistic about science. He said, "As long as we have a free society in which the human intellect can operate at its best, we will come out on top victorious, triumphant, and full of hope for a better and finer future."[93]

Only a few of Laurence's journalistic descendants have recognized the limitations of this sunny view of science and technology. One was Natalie Angier, a successor of Laurence's at the *Times*. "We science journalists, perhaps more than any other class of reporters, too often serve as perky cheerleaders for our subject and our sources," she wrote in 1991. "Maybe that is because most of us really do love science; or maybe we are so worried that the rest of the world does not that we feel obliged to bring out all the bells, whistles and bullhorns. But in any case we sometimes end up writing copy that sounds like an unvarnished press release."[94] William J. Broad, a longtime science journalist at the *Times*, wrote in 2021 that Laurence was "not only an apologist for the American military but also a serial defier of journalism's mores." But Broad spoke only for himself; even in his lengthy examination of Laurence's work occupying a prominent place in the paper's Science section, no senior *Times* official was quoted as criticizing Laurence's behavior.[95]

Similarly, when Laurence died in 1977, the *Times* would not publicly disavow the professional accomplishments of the award-winning reporter who had garnered it so much positive attention even while alienating reporters and editors inside the organization. In its front-page obituary of Laurence, the *Times* reprised his childhood in Lithuania, his tutoring business at Harvard, and the arc of his journalistic career, including the uncovering of atomic research and his work for the Manhattan Project. The *Times* also scaled back its earlier claim that Laurence had been the first full-time daily science reporter, stating in the obituary that Laurence had been "one of the country's first full-time science reporters."[96] Five days later the *Times* editorialized that Laurence "transmitted to a whole era not only information and analysis but excitement about the most complex ideas, the kind science had previously earned only from remote, romantic expeditions. It is the thrill of discovery inside the mind that Bill Laurence captured."[97]

The Associated Press's obituary called Laurence "the only journalist to witness the dawn of the nuclear age" and noted his publication of his article on cancer in the Spanish medical journal. "He recently underwent a stomach operation and then suffered a blood clot in the brain," the wire service reported.[98] In an obituary in the NASW newsletter, Earl Ubell, health and science editor of WCBS-TV in New York and former science editor for the *New York Herald Tribune*, recalled the experience of being Laurence's journalistic competitor. "He was the rabbit after whom the young dogs ran," Ubell wrote. "At 70, his glee at having scooped the rest of us matched that of any cub reporter. Yet, he never held off sharing with us his prodigious knowledge—lecturing us kindly with a mixed Harvard-Russian accent piped nasally through a broken nose on the details of pituitary function, Hilbert's problems, or the carbon cycle of photosynthesis."[99]

Had Laurence died in the United States, he undoubtedly would have been eulogized at a funeral attracting journalists, scientists, and government leaders. But Florence had him buried quietly and without delay in the Palma de Mallorca Municipal Cemetery.[100] "It was very simple—that's the way he would have wanted it," Florence Laurence told a *Times* staffer.[101]

And so, William L. Laurence—the maestro of drama who lived on the front page of one of the most influential newspapers in the world—slipped quietly into history.

Acknowledgments

I am indebted to the staff members at the dozens of archives listed in the notes, who unfailingly retrieved boxes of materials, some not opened in years, for me to plumb, or in other cases promptly responded to remote requests for documents. In particular, Jeff Flannery and his colleagues at the Manuscript Reading Room at the Library of Congress and Tal Nadan and her colleagues at the Brooke Russell Astor Reading Room for Rare Books and Manuscripts at the New York Public Library receive my gratitude.

Others who offered me invaluable support, encouragement, and help include Walter Rankin, Michael Canter, Trey Sullivan, Jessica Vanderhoff, Marcel LaFollette, Becky Bailey, Bill Broad, Alex Wellerstein, Carole Sargent, Bibi Gaston, Sonia Taitz, and James Nolan. Marlin Rout provided an insightful tour of the Trinity Site.

The project would not have come to a successful conclusion without the patient guidance of Michael McGandy, my editor at Cornell University Press and the editorial director of the Three Hills imprint. Others at the Press who were vital to this book include Clare Jones, who masterfully prepared the manuscript and art; Karen Laun, whose keen editor's eye kept me out of more than one ditch; Susan Specter and Karen Hwa, who shared duties as production editor; Scott Levine and his colleagues, who designed the book cover; Lori Rider and Todd Rider, who assisted with copyediting; and Sarah Noell, our talented publicist. Thanks to Colleen Dunham for producing the book's index.

As in all aspects of my life, I relied on the loving support of my spouse, Terri, and my children, Emily and Matthew. Thanks. I love you with my whole heart.

Notes

Introduction

1. Hillier Krieghbaum, "American Newspaper Reporting of Science News," *Kansas State College Bulletin* 25, no. 5 (1941): 45, 65.

2. William L. Laurence, "Theory Based on Radiation Which Disputes Einstein's Offered by Dartmouth Savant," *World,* January 1, 1930; William L. Laurence, "Restores Laws in Nature, Gives New Intelligibility," *World,* January 5, 1930.

1. The Second Coming of Prometheus

1. William L. Laurence, "Energy Multiplied 200,000,000 Times," *New York Times,* April 27, 1935.

2. "Seven Days' Survey," *Commonweal,* May 10, 1935, 43.

3. Thomas R. Henry, "Power of New Atomic Blast Greatest Achieved on Earth," *Evening Star* (Washington, DC), January 28, 1939.

4. "200 Million Volts of Energy Created by Atom Explosions," *San Francisco Chronicle,* January 30, 1939; Luis W. Alvarez, *Alvarez: Adventures of a Physicist* (New York: Basic Books, 1987), 72.

5. Laurence interview, 256–57.

6. "Atom Explosion Frees 200,000,000 Volts; New Physics Phenomenon Credited to Hahn," *New York Times,* January 29, 1939; Klaus Hoffman, *Otto Hahn: Achievement and Responsibility*, trans. J. Michael Cole (New York: Springer-Verlag, 2001), 129.

7. William L. Laurence, *Men and Atoms: The Discovery, the Uses, and the Future of Atomic Energy* (New York: Simon & Schuster, 1959), 6–7.

8. Laurence, *Men and Atoms*, 9.

9. John J. O'Neill, *Almighty Atom: The Real Story of Atomic Energy* (New York: Ives Washburn, 1945), 45.

10. William L. Laurence, "Vast Power Source in Atomic Energy Opened by Science," *New York Times*, May 5, 1940.

11. William L. Laurence, "First with Atomic Story," *Saturday Review of Literature*, December 20, 1947; John J. O'Neill, "'Your Newspaper,'" *Saturday Review of Literature*, November 22, 1947.

12. "William L. Laurence," Sulzberger Papers, box 11, folder 13.

13. "Secret of Atomic Power Revealed by 'U' Professor," *Minneapolis Sunday Tribune*, May 5, 1940.

14. Raymond Moscowitz, *Stuffy: The Life of Newspaper Pioneer Basil "Stuffy" Walters* (Ames: Iowa State Press, 1982); "U. of M. Man's Discovery Opens Vast Power Source," *Minneapolis Star-Journal*, May 5, 1940, emphasis in original.

15. "Vast Atomic Power Possible If Enough Uranium Is Isolated," *Newsweek*, May 13, 1940, 41.

16. *Congressional Record* (76th Cong., 3rd Sess.), August 15, 1940, 10392.

17. Waldemar Kaempffert, "Science in the News," *New York Times*, May 12, 1940.

18. Waldemar Kaempffert, "Scientific Men and the Newspapers," *Science* 81 (1935): 640.

19. Oral history of Turner Catledge (1979), Times oral histories, box 2, folder 2, 200.

20. Interview of Henry A. Barton by Charles Weiner, March 16, 1970, https://www.aip.org/history-programs/niels-bohr-library/oral-histories/4494-2.

21. Barton interview by Weiner.

22. Advisory Committee on Uranium, letter to General Edwin M. Watson, May 1940, Nier Papers, box 2, folder 34.

23. "U-235 Power Held of No War Use Now," *New York Times*, May 6, 1940.

24. Correspondence from George B. Pegram to Lyman J. Briggs, May 6, 1940, Nier Papers, box 2, folder 34.

25. Correspondence from George B. Pegram to Hans A. Bethe, May 14, 1940, Bethe Papers, box 89, folder 9–11.

26. Correspondence from J. R. Dunning to Alfred O. Nier, May 9, 1940, Nier Papers, box 2, folder 34.

27. Howard W. Blakeslee, "Security Force of 485 Went to Great Lengths to Insure Atom Bomb Security," *Joplin (MO) Globe*, October 16, 1945.

28. James A. Michener, *The World Is My Home: A Memoir* (New York: Random House, 1992), 253–55.

29. Patti Clayton Becker, *Books and Libraries in American Society during World War II: Weapons in the War of Ideas* (New York: Routledge, 2005).

30. David Holloway, *Stalin and the Bomb: The Soviet Union and Atomic Energy, 1939–1956* (New Haven, CT: Yale University Press, 1994), 61.

31. Edward Wysocki, "The Creation of Heinlein's 'Solution Unsatisfactory,'" in *Practicing Science Fiction: Critical Essays on Writing, Reading and Teaching the Genre*, ed. Karen Hellekson (Jefferson, NC: McFarland, 2010), 74–86; Arthur McCann, "Shhhhh! Don't Mention It," *Astounding Science-Fiction*, August 1940, 104.

32. Peter D. Smith, *Doomsday Men: The Real Dr. Strangelove and the Dream of the Superweapon* (London: Penguin, 2008).

33. Correspondence from William L. Laurence to Karl T. Compton, June 18, 1940, Compton Papers, box 133, folder 2.

34. Correspondence from Philip M. Morse to K. T. Compton, June 24, 1940, Morse Papers, box 8, folder "L—General."

35. Correspondence from John J. Rowlands to William L. Laurence, June 25, 1940, Compton Papers, box 133, folder 2.

36. Correspondence from Karl T. Compton to William L. Laurence, July 5, 1940, Compton Papers, box 133, folder 2.

37. Correspondence from William L. Laurence to Philip M. Morse, July 5, 1940, Morse Papers, box 8, folder "L—General."

38. Correspondence from Philip M. Morse to William L. Laurence, July 10, 1940, Morse Papers, box 8, folder "L—General."

39. Correspondence from Edwin McMillan to William L. Laurence, July 30, 1940, Lawrence Papers, carton 23, folder 34, reel 35.

40. William L. Laurence, "The Atom Gives Up," *Saturday Evening Post*, September 7, 1940, 13.

41. Laurence, "Atom Gives Up," 60.

42. Laurence, "Vast Power Source in Atomic Energy," 51.

43. Ruth Lewin Sime, *Lise Meitner: A Life in Physics* (Berkeley: University of California Press, 1996), 315.

44. "Origins," *Time*, August 20, 1945, 33.

45. "Meitner, Lise," in *Current Biography* (New York: H. W. Wilson, 1945), 393–95.

46. Ulla Bolin, "Dr. Lise Meitner Deplores Use of Atom Bomb," *New York Herald Tribune*, August 12, 1945.

47. William L. Laurence, "Discovery of Atomic Power International Drama," *Washington Post*, August 12, 1945.

48. Laurence, "Atom Gives Up," 62.

49. Laurence, "Atom Gives Up," 62.

50. Laurence, "Atom Gives Up," 12, 63.

2. On the Army's Payroll

1. Vincent Kiernan, *Embargoed Science* (Urbana: University of Illinois Press, 2006), 176.

2. National Research Council, Division of Medical Sciences, "Committee on Information. Minutes of Meeting, August 26, 1940," Subcommittee on Publicity Minutes, microfiche, NAS archives.

3. Correspondence from David Dietz to Lewis H. Weed, September 3, 1940, folder "Com on Information, Subcoms: Publicity, 1940–1946," NAS archives.

4. Correspondence from Howard W. Blakeslee to Lewis H. Weed, August 29, 1940, folder "Com on Information, Subcoms: Publicity, 1940–1946," NAS archives.

5. Correspondence from Lewis H. Weed to Morris Fishbein, September 25, 1940; correspondence from Morris Fishbein to Lewis H. Weed, September 16, 1940; correspondence from Morris Fishbein to Lewis H. Weed, September 27, 1940; correspondence from Lewis H. Weed to Morris Fishbein, October 22, 1940, all in folder "Com on Information, Subcoms: Publicity, 1940–1946," NAS archives.

6. Correspondence from Lewis H. Weed to Morris Fishbein, October 28, 1940, folder "Com on Information, Subcoms: Publicity, 1940–1946," NAS archives.

7. "Medical Preparedness," *Journal of the American Medical Association* 115, no. 19 (November 9, 1940), 1640–43.

8. National Research Council, Division of Medical Sciences, "Subcommittee on Publicity, Minutes of Meeting, December 11, 1940," Science Service Records, box 250, folder 3, 4.

9. Correspondence from David Dietz to Wallace Werble, November 22, 1940, Science Service Records, box 250, folder 3.

10. National Research Council, Division of Medical Sciences, Committee on Information and Subcommittee on Publicity, "Minutes of the Thirteenth Meeting. June 28 and 29, 1943," "Subcommittee on Publicity Minutes," (microfiche), 1, NAS archives.

11. J. G. Frierson, "The Yellow Fever Vaccine: A History," *Yale Journal of Biology and Medicine* 83, no. 2 (2010): 77–85.

12. "Colonel Dies of Jaundice," *New York Times,* August 4, 1942, 10; "Army Curbs Jaundice Traced to a Serum," *New York Times,* July 25, 1942; Waldemar Kaempffert, "Jaundice in Army," *New York Times,* October 18, 1942; "Jaundice Attacks Soldiers," *New York Times,* June 30, 1942.

13. "A Grievous Error," *Chicago Daily Tribune,* July 27, 1942; "The Yellow Fever Inoculation," *Chicago Daily Tribune,* July 30, 1942.

14. Morris Fishbein, "Memorandum on Publicity," folder "Com on Information, Subcoms: Publicity, 1940–1946," 4, NAS archives.

15. Morris Fishbein, "Memorandum on Publicity," 5–6.

16. Robert D. Potter, "Dear Editor," *Newsletter of the National Association of Science Writers,* March 1962, 40.

17. John R. Paul and Horace T. Gardner, "Viral Hepatitis," in *Communicable Diseases,* vol. 5 (Washington, DC: Office of the Surgeon General, Department of the Army, 1955), 411–62.

18. Correspondence from Norman T. Kirk and George F. Lull to William L. Laurence, February 26, 1944, Laurence's official personnel folder.

19. Correspondence from George F. Lull to Charles Merz, March 22, 1944, Laurence's official personnel folder.

20. Jerome Karabel, *The Chosen: The Hidden History of Admission and Exclusion at Harvard, Yale and Princeton* (Boston: Houghton Mifflin, 2009).

21. "Middlesex Court Matters," *Cambridge (MA) Chronicle,* September 4, 1915, 10.

22. Correspondence from Morris Fishbein to Brig. Gen. Fred Rankin, March 22, 1944, folder "Com on Information, Subcoms: Publicity, 1940–1946," NAS archives; Division of Medical Sciences of the National Research Council, Subcommittee on Publicity, "Minutes of Meeting, 18 March 1944," "Publicity—Bulletin, numbered abstracts and reports" (microfiche), NAS archives.

23. John W. Martyn, Memorandum to the Surgeon General, April 26, 1944, Laurence's official personnel folder.

24. 201 File, July 22, 1946, Laurence's official personnel folder.

25. "Army Appoints Civil Consultants," *Salt Lake Tribune,* April 30, 1944; "Science Writers Aid Army Medical Study," *Editor and Publisher,* May 20, 1944, 50; "Five Writers to Aid Surgeon-General," *Richmond (VA) Times-Dispatch,* April 30, 1944.

26. Robert D. Potter, "History of NASW," NASW archives, box 4, folder 27, 38.

27. War Department, "Individual Earnings Record, William L. Laurence," card no. 3943, Laurence's official personnel folder.

28. Potter, "Dear Editor," 41.

29. Potter, "History of NASW"; correspondence from Lewis H. Weed to David Dietz, June 15, 1946, Dietz Papers, box 16, folder "Office of Surgeon General."

30. Selective Service System Records, record U778.

31. "The Oregon Code of Ethics for Journalism Adopted at the Oregon Newspaper Conference, 1922," *Annals of the American Academy of Political and Social Science* 101 (1922): 285.

32. "Sigma Delta Chi's New Code of Ethics," Society of Professional Journalists, http://spjnetwork.org/quill2/codedcontroversey/ethics-code-1926.pdf (accessed May 29, 2020).

33. "Sigma Delta Chi's New Code of Ethics"; War Department, Bureau of Public Relations, Press Branch, "Soldier Continues to Fight with Shell Fragment Lodged on Heart," Dietz Papers, box 21, folder "S.G.O.-Travel Orders."

34. "Shell Fragment Is Taken from Spot by Soldier's Heart," *Lubbock (TX) Morning Avalanche*, July 11, 1944; Frank Carey, "Yank Fights through Campaign with Shell Fragment in Heart," *Charleston (WV) Gazette*, July 10, 1944.

35. "Science Bolsters GIs' Tired Dogs," *Charleston (WV) Gazette*, July 25, 1944; " 'March Fractures' Result from Intensive Training," Dietz Papers, box 21, folder "S.G.O.-Travel Orders."

36. "Malaria Epidemic Danger in United States Is Remote," Dietz Papers, box 21, folder "S.G.O.-Travel Orders"; Frank Carey, "Atabrine Held Malaria Cure," *Arizona Republic* (Phoenix), September 7, 1944.

37. "Fewer Leg and Arm Wounds in Present War than in Civil War and World War," "New Oil Treatment Reduces Spread of Bacteria and Viruses," and "Patients on Hospital Ships Now Get Whole Milk," all in Dietz Papers, box 21, folder "S.G.O.-Travel Orders."

38. "Army Veterinary Corps" and "Army's Epidemiology Board Finds New Methods to Conquer Disease," both in Dietz Papers, box 21, folder "S.G.O.-Travel Orders."

39. Ernst Chain et al., "Penicillin as a Chemotherapeutic Agent," *Lancet* 236, no. 6104 (1940): 226–28.

40. William L. Laurence, "Journalistic Aspects of Science Writing," in *Writing in Industry*, vol. 1, ed. Siegfried Mandel (Brooklyn: Polytechnic Press, 1959), 105.

41. Steven M. Spencer, "Germ Killer Found in Common Mold," *Philadelphia Evening Bulletin*, May 5, 1941.

42. William L. Laurence, " 'Giant' Germicide Yielded by Mold," *New York Times*, May 6, 1941.

43. "From the Archives," *Science, Technology, and Human Values* 10, no. 2 (Spring 1985): 24–27.

44. William L. Laurence, "More Penicillin," *New York Times*, August 1, 1943.

45. Robert D. Potter, "Letter to the Editor," *Newsletter of the National Association of Science Writers*, March 1962, 41.

46. "Malaria Remedy Tested in Prison," *New York Times*, July 23, 1944.

47. National Research Council, Division of Medical Sciences, Committee on Information and Subcommittee on Publicity, "Minutes of Meeting—1 March 1945," "Publicity—Bulletin, numbered abstracts and reports", microfiche, NAS archives.

48. William L. Laurence, "New Drugs to Combat Malaria Are Tested in Prisons for Army," *New York Times*, March 5, 1945.

49. "Prison Malaria," *Life*, June 4, 1945, 43; "200 Convicts in U.S. Used in Tests for Malaria Cure," *Mediterranean Stars and Stripes*, March 6, 1945; Jon M. Harkness, "Nuremberg and the Issue of Wartime Experiments on US Prisoners: The Green Committee," *Journal of the American Medical Association* 276, no. 20 (November 27, 1996): 1672–75.

50. James H. Stack, "James T. Grady Award," *Newsletter of the National Association of Science Writers*, June 1958, 23.

51. Stack, "James T. Grady Award."

52. William L. Laurence, “Eyewitness to a Later Day Genesis,” Schuster Papers, box 274, folder “William L. Laurence,” 9.

3. Magnetic Current

1. Gerald Holton, “Subelectrons, Presuppositions, and the Millikan-Ehrenhaft Dispute,” *Historical Studies in the Physical Sciences* 9 (1978): 161–224; “Moving Power of Light,” *Los Angeles Times,* December 14, 1919; “Doubt Electron Theory,” *New York Times,* September 24, 1926.

2. Correspondence from Warren Weaver to Emil Witochi, September 7, 1939, and correspondence from Emil Witochi to Warren Weaver, September 5, 1939, both in Rockefeller Foundation files, Series 705D, box 3, folder 26; Helge Kragh, “The Concept of the Monopole: A Historical and Analytical Case-Study,” *Studies in History and Philosophy of Science Part A* 12, no. 2 (1981): 141–72.

3. “Reported from the Fields of Research,” *New York Times,* March 3, 1940; “Science News,” *Science* 91, no. 2357 (March 1, 1940): 10–12, 14; “New Explanation Evolved for Sun's Flaming Corona,” *Science News-Letter* 37, no. 9 (March 2, 1940): 131–32.

4. P. A. M. Dirac, *Ehrenhaft, the Subelectron and the Quark* (New York: Academic Press, 1977).

5. Correspondence from H. M. Miller to Alice Lili Rona, March 26, 1941, and correspondence from Alice Lili Rona to Raymond B. Fostick, March 21, 1941, Rockefeller Foundation files, Series 705D, box 3, folder 26; Felix Ehrenhaft and L. Banet, “Magnetization of Matter by Light,” *Nature* 147 (1941): 297; “Reported from the Field of Research,” *New York Times,* April 13, 1941; G. B. Lal, “Science News,” *Washington Post,* January 26, 1941.

6. Correspondence from William L. Laurence to Mr. Sulzberger, April 13, 1944, Rockefeller Foundation files, Series 705D, box 3, folder 27; Ehrenhaft and Banet, “Magnetization of Matter by Light,” 297.

7. Laurence letter to Sulzberger, April 13, 1944. Laurence's recollection of the second paper also seems to be off the mark; the quotation he cites appears in the earlier *Science* paper and not the second.

8. Laurence letter to Sulzberger, April 13, 1944. A biographer of Ehrenhaft independently confirms Laurence's presence at the Columbia lecture: Joseph Braunbeck, *Der andere Physiker: Das Leben von Felix Ehrenhaft* (Vienna: Technisches Museum Wien, 2003).

9. Laurence letter to Sulzberger, April 13, 1944.

10. Lilly Rona, Journal, January 11, 1944, Ehrenhaft Papers, box 1, folder 4.

11. Laurence letter to Sulzberger, April 13, 1944.

12. William L. Laurence, “Proof Offered of Existence of Pure Magnetic Current,” *New York Times,* January 16, 1944.

13. Laurence, “Proof Offered of Existence of Pure Magnetic Current.”

14. “New Magnetic Currents Proof Reported Found,” *Los Angeles Times,* January 16, 1944.

15. Lilly Rona, Journal, January 15, 1944, Ehrenhaft Papers, box 1, folder 4.

16. “Ehrenhaft,” *New Yorker,* January 29, 1944, 16.

17. “Magnetism in Harness?” *Time,* January 24, 1944, 36.

18. Alfred Werner, “An Epochal Discovery,” *Liberal Judaism,* February 1944, 46.

19. Lilly Rona, Journal, February 16, 1944, Ehrenhaft Papers, box 1, folder 5.

20. Correspondence from Gordon Ferrie Hull to Professor [Raymond T.] Birge, November 27, 1946, Birge Papers, box 15.

21. Correspondence from Felix Ehrenhaft to Raymond D. Fosdick, February 28, 1944, and correspondence from Warren Weaver to Felix Ehrenhaft, March 6, 1944, both in Rockefeller Foundation files, Series 705D, box 3, folder 27.

22. Waldemar Kaempffert, "Is There a Magnetic Current?" *New York Times*, February 27, 1944; James T. Kendall, "'Magnetic' Current," *Nature* 153, no. 3875 (1944): 157–58.

23. Lilly Rona, Journal, March 1, 1944, Ehrenhaft Papers, box 1, folder 5.

24. "Old Rule in Physics Called 'Fairy Tale,'" *New York Times*, March 22, 1944.

25. Roger W. Stuart, "Wife Helps Scientist Prove Belief Wrong on Magnetic Current," *Pittsburgh Press*, April 4, 1944; Roger W. Stuart, "Wife Helps Scientist Prove 700-Year Belief Is Wrong on Magnetic Current," *New York World-Telegram*, March 16, 1944.

26. Correspondence from Warren Weaver to H. A. Barton, March 17, 1944, Rockefeller Foundation files, Series 705D, box 3, folder 27.

27. Correspondence from H. A. Barton to Warren Weaver, March 31, 1944, Rockefeller Foundation files, Series 705D, box 3, folder 27.

28. Laurence letter to Sulzberger, April 13, 1944; correspondence from Arthur Hays Sulzberger to Warren Weaver, April 18, 1944, Rockefeller Foundation files, Series 705D, box 3, folder 27.

29. Correspondence from Warren Weaver to Arthur Hays Sulzberger, April 17, 1944, Sulzberger Papers, box 42, folder 1.

30. Correspondence from George B. Pegram to Warren Weaver, April 21, 1944, Rockefeller Foundation files, Series 705D, box 3, folder 27.

31. Waldemar Kaempffert, "Ehrenhaft's Theory of a Magnetic Current Is Challenge to Physical Scientists," *New York Times*, April 23, 1944.

32. "Offers New Tests on Magnetic Flow," *New York Times*, April 29, 1944.

33. Correspondence from Arthur Hays Sulzberger to Warren Weaver, May 1, 1944, Rockefeller Foundation files, Series 705D, box 3, folder 27.

34. Weaver letter to Sulzberger, May 1, 1944.

35. Correspondence from Raymond E. Fosdick to Mrs. Felix Ehrenhaft, April 19, 1944; correspondence from Lilly Rona-Ehrenhaft to Raymond D. Fosdick, April 10, 1944; and Warren Weaver, Inter-Office Memorandum Subject: Lilly Rona-Ehrenhaft to RBF—letter of April 10, 1944, dated April 12, 1944, all in Rockefeller Foundation files, Series 705D, box 3, folder 27.

36. John W. Campbell Jr., "Beachhead for Science," *Astounding Science-Fiction*, May 1944, 103.

37. Alden P. Armagnac, "Magic with Magnetism," *Popular Science*, June 1944, 130.

38. Carl Olsson, "Your Servant, the Magnet," *Illustrated*, June 10, 1944, 12.

39. William L. Laurence, "Magnetic Motor Maps 'Third Force,'" *New York Times*, June 25, 1944.

40. William L. Laurence, "A Physicist Supports His Hypothesis That Magnetic Currents Flow through Universe," *New York Times*, July 30, 1944.

41. Correspondence from William L. Laurence to William F. G. Swann, August 11, 1944, and correspondence from William F. G. Swann to William L. Laurence, August 4, 1944, both in Swann Papers, box 28, folder "Laurence, William L. 1944–1957."

42. Correspondence from Lilly Rona to William L. Laurence, August 3, 1944, Ehrenhaft Papers, box 2, folder 5.

43. Lilly Rona, Journal, September 24, 1944, Ehrenhaft Papers, folder 5, box 1.

44. Gildo Magalhaes Santos, "A Debate on Magnetic Current: The Troubled Einstein–Ehrenhaft Correspondence," *British Journal for the History of Science* 44, no. 3 (2011): 371–400.

45. William L. Laurence, "Expert Says Light Can Rotate Matter," *New York Times*, January 20, 1945; annotated copy of Laurence's article in Rockefeller Foundation files, box 3, folder 27.

46. "What Is Light?" *Time*, February 5, 1945, 76; Howard W. Blakeslee, "New Concept of Light's Motion Told Scientists," *Los Angeles Times*, January 20, 1945; "Tests Show Light Rotates Solids, May Explain Earth's Spin," *Washington Post*, January 20, 1945; "Scientist Says Light Travels Spiral Path," *Baltimore Sun*, January 20, 1945.

47. Harold S. Renne, "Magnetic Current," *Radio News*, April 1945, 22.

48. "Ehrenhaft Discovery Confirmed by New Experiments," *Popular Science*, April 1945, 208.

49. Howard W. Blakeslee, "Viennese Physicist Claims Sunshine Carries Strong Magnetism Charge," *Amarillo (TX) Daily News*, June 22, 1945; "Reaching for the Sun," *Newsweek*, July 2, 1945.

50. Correspondence from F. Ehrenhaft to Rockefeller Foundation, February 20, 1947, and correspondence from Warren Weaver to F. Ehrenhaft, March 4, 1947, both in Rockefeller Foundation files, Series 705D, box 3, folder 27.

51. C. F. Bikel, Memorandum to Director of Intelligence, July 22, 1948, Army Staff records, entry 134B, file XA010075.

52. Santos, "Debate on Magnetic Current."

53. "Felix Ehrenhaft, 73, Physicist, Professor," *New York Times*, March 6, 1952.

54. Correspondence from S. A. Goudsmit to William L. Laurence, August 1, 1958, Goudsmit Papers, folder 135, box 14.

55. Correspondence from Raymond T. Birge to Gordon F. Hull, December 7, 1946, Hull Papers, box 9, folder 33.

56. Warren Weaver, 1951 diary entry for March 14, 1951, pp. 52–53, Rockefeller Foundation files, Officers' Diaries, RG 12, S–Z.

4. Atomland-on-Mars

1. Richard D. Easton and Eric F. Frazier, *GPS Declassified: From Smart Bombs to Smartphones* (Lincoln, NE: Potomac Books, 2013); William L. Laurence, "'Cosmic Pendulum' for Clock Planned," *New York Times*, January 21, 1945, 34.

2. William Leonard Laurence, interview by Scott Bruns, Columbia Center for Oral History, Columbia University, New York, 278.

3. Laurence interview by Bruns, 282.

4. S. J. Monchak, "Laurence Relates His Role on Atomic Bomb Project," *Editor and Publisher*, September 22, 1945, 9.

5. William L. Laurence, *Men and Atoms: The Discovery, the Uses, and the Future of Atomic Energy* (New York: Simon & Schuster, 1959), 95–96.

6. Michael S. Sweeney, *Secrets of Victory: The Office of Censorship and the American Press and Radio in World War II* (Chapel Hill: University of North Carolina Press, 2001), 204.

7. Patrick Scott Washburn, "The Office of Censorship's Attempt to Control Press Coverage of the Atomic Bomb during World War II," *Journalism Monographs* 120 (1990).

8. L. R. Groves, *Now It Can Be Told* (New York: Harper & Row, 1962).

9. Groves, *Now It Can Be Told*, 325.

10. Michael D. Gordin, *Five Days in August: How World War II Became a Nuclear War* (Princeton, NJ: Princeton University Press, 2007), 110.

11. William L. Laurence, "The Greatest Story," in *How I Got That Story*, ed. David Brown and Richard Bruner (New York: E. P. Dutton, 1967), 190–91.

12. Correspondence from L. R. Groves to Arthur H. Sulzberger, September 12, 1945, Groves Papers, entry 4, box 1, folder "MED—Correspondence (Misc)."

13. Groves, *Now It Can Be Told*, 326.

14. Laurence, *Men and Atoms*, 96.

15. Memorandum from L. R. Groves to Colonel K. D. Nichols, May 9, 1945, Groves Papers, entry 4, box 1, folder "MED—Correspondence (Misc)."

16. Robert S. Norris, *Racing for the Bomb: General Leslie R. Groves, the Manhattan Project's Indispensable Man* (South Royalton, VT: Steerforth Press, 2002).

17. Arthur Gelb, *City Room* (New York: Putnam, 2003), 95.

18. Laurence interview by Bruns, 286.

19. Laurence interview by Bruns, 288.

20. George O. Robinson, *The Oak Ridge Story: The Saga of a People Who Share in History* (Kingsport, TN: Southern Publishers, 1950), 102.

21. Leonard Lyons, "The Lyon's Den," *Morning Advocate*, February 26, 1949.

22. Richard G. Hewlett and Oscar E. Anderson, *The New World, 1939/1946*, vol. 1 (University Park: Pennsylvania State University Press, 1962).

23. Lansing Lamont, *Day of Trinity* (New York: Atheneum, 1985), 107.

24. Ann Jacob Stocker, "A Study of the Life and Work of William L. Laurence as Related to His Contributions in Science Journalism" (master's thesis, Syracuse University, 1966), 70.

25. "TRINITY Test (at Alamogordo, July 16, 1945)," Manhattan District correspondence, roll 1, target 5, folder 4.

26. Stocker, "Study of the Life and Work," 70.

27. "TRINITY Test."

28. Karl Grossman, *The Wrong Stuff: The Space Program's Nuclear Threat to Our Planet* (Monroe, ME: Common Courage Press, 1997), 176.

29. Corinne Browne and Robert L. Munroe, *Time Bomb: Understanding the Threat of Nuclear Power* (New York: Morrow, 1981), 37.

30. Gelb, *City Room*, 101.

31. Alex Wellerstein, *Restricted Data: The History of Nuclear Secrecy in the United States* (Chicago: University of Chicago Press, 2021), 111; Peter N. Kirstein, "Hiroshima and Spinning the Atom: America, Britain, and Canada Proclaim the Nuclear Age, 6 August 1945," *Historian* 71, no. 4 (2009): 812.

32. William L. Laurence, Tentative Draft of Radio Address by President Truman to Be Delivered after the Successful Use of the Atomic Bomb over Japan, Harrison-Bundy Files, roll 6, file 74, 2.

33. Correspondence from James B. Conant to Vannevar Bush, May 18, 1945, Bush-Conant Files, roll 4, target 1.

34. Hewlett and Anderson, *New World*, 354.

35. Statement by the President of the United States, August 6, 1945, Manhattan District History, roll 1.

36. Memorandum from William A. Consodine to Mr. Page, June 19, 1945, Manhattan District Records, entry 5, box 31, file "000.71 (Releasing Information)."

37. Laurence interview by Bruns, 395.

38. Laurence, "Greatest Story," 190.

39. Groves, *Now It Can Be Told*, 327.

40. Monchak, "Laurence Relates His Role."

41. Laurence, *Men and Atoms*, 97.

42. J. F. Sally, Production Office Diary, July 17, 1944 through January 24, 1946, accession no. RL-1–345586; Franklin T. Matthias, "Diary—Colonel FT Matthias—1945,

Hanford Engineer Works," accession no. RL-1–331968, both on https://www.osti.gov/opennet/ (accessed May 29, 2020).

43. William L. Laurence, *Dawn over Zero: The Story of the Atomic Bomb* (New York: Knopf, 1946), 116–17.

44. Jennet Conant, *109 East Palace: Robert Oppenheimer and the Secret City of Los Alamos* (New York: Simon & Schuster, 2005), 292.

45. Stocker, "Study of the Life and Work," 69–70.

46. Laurence, *Dawn over Zero*, 116.

47. William L. Laurence, *The Hell Bomb* (New York: Knopf, 1951), 17–18.

48. Glenn T. Seaborg, *History of Met Lab Section C-I, May 1945 to May 1946*, Pub. 112, vol. 4, Lawrence Berkeley Laboratory, University of California, Berkeley, 83.

49. Consodine letter to Mr. Page.

50. "An Account of the Discovery and Early Study of Uranium 233," June 30, 1945, Seaborg Papers, box 172, folder 4, 2.

51. Laurence interview by Bruns, 297.

52. Laurence, *Men and Atoms*, 104.

53. "Is Your Town a Choice A-Bomb Target?," *Saturday Evening Post*, November 6, 1948, 6.

54. Laurence, *Men and Atoms*.

55. Monchak, "Laurence Relates His Role," 9.

56. Norris, *Racing for the Bomb*.

57. Beverly Ann Deepe Keever, "TOP SECRET: Censoring the First Rough Drafts of Atomic Bomb History," *Media History* 14, no. 2 (2008): 193.

58. George O. Robinson, "Suggestions Regarding Release of Story," June 6, 1945, Manhattan District Records, entry 5, box 31, folder "000.71 (Releasing Information)."

59. Laurence, *Men and Atoms*.

60. Meyer Berger, *The Story of the New York Times, 1851–1951* (New York: Simon & Schuster, 1951), 512.

61. Berger, *Story of the New York Times*.

62. William L. Laurence, "Laurence Reveals 'Atomic' Letter to James; Had Story of 'New Gas' If Bomb Misfired," *Timesweek*, June 11, 1947, 1, Catledge Papers, series IIA, folder "Timesweek 1947," 4.

63. Berger, *Story of the New York Times*, 514.

5. Trinity, Hiroshima, and Nagasaki

1. William L. Laurence, "Now We Are All Sons-of-Bitches," *Science News*, July 11, 1970, 39.

2. Lansing Lamont, interview with William L. Laurence, Lamont Papers, folder 2, box 1, 1.

3. William Leonard Laurence, interview with Scott Bruns, Columbia Center for Oral History, Columbia University, New York; US Department of Energy, *United States Nuclear Tests, July 1945 through September 1992* (Las Vegas: US Department of Energy Nevada Operations Office), https://www.nnss.gov/docs/docs_LibraryPublications/DOE_NV-209_Rev16.pdf (accessed May 29, 2020).

4. Lansing Lamont, *Day of Trinity* (New York: Atheneum, 1985); Stephane Groueff, *Manhattan Project: The Untold Story of the Making of the Atomic Bomb* (Boston: Little, Brown, 1967); correspondence from J. Chadwick to Lansing Lamont, September 9, 1964, Lamont papers, box 1, folder "Correspondence concerning *Day of Trinity* by Lansing Lamont (New York: Atheneum, 1965)."

5. Lamont interview with Laurence, 6.
6. Lamont, *Day of Trinity*.
7. Lamont interview with Laurence, 7.
8. Laurence interview with Bruns, 315.
9. Lamont interview with Laurence, 9; Joseph W. Kennedy, Diary, Sunday, July 15 and Monday, July 16, 1945, Seaborg Papers, box 172, folder 6.
10. Los Alamos Scientific Laboratory, *Los Alamos: Beginning of an Era, 1943–1945* (Los Alamos, NM: LASL, 1967), 50.
11. Lamont, *Day of Trinity*, 229.
12. William L. Laurence, "Drama of the Atomic Bomb Found Climax in July 16 Test," *New York Times*, September 26, 1945.
13. Radio interview with William L. Laurence, June 22, 1946, Recorded Sound Research Center, recording ICD 26504.
14. Cynthia C. Kelly, ed., *The Manhattan Project: The Birth of the Atomic Bomb in the Words of its Creators, Eyewitnesses, and Historians* (New York: Black Dog & Leventhal, 2007), 310.
15. Victor Frederick Weisskopf, *The Joy of Insight: Passions of a Physicist* (New York: Basic Books, 1991), 151.
16. Groueff, *Manhattan Project*, 356.
17. Interview of Dr. Robert E. Marshak by Charles Weiner on September 19, 1970, American Institute of Physics, https://www.aip.org/history-programs/niels-bohr-library/oral-histories/4494-2 (accessed May 29, 2020).
18. Jeremy Bernstein, *Nuclear Weapons: What You Need to Know* (Cambridge: Cambridge University Press, 2008), 181.
19. Laurence, "Drama of the Atomic Bomb," 16.
20. Laurence, "Drama of the Atomic Bomb," 16.
21. Laurence, "Drama of the Atomic Bomb," 16.
22. Lamont, "Notes," Lamont Papers, box 1, folder 2, 141.
23. William L. Laurence, *Men and Atoms: The Discovery, the Uses, and the Future of Atomic Energy* (New York: Simon & Schuster, 1959), 118.
24. James A. Hijiya, "The 'Gita' of J. Robert Oppenheimer," *Proceedings of the American Philosophical Society* 144, no. 2 (June 2000): 123–24, n. 3.
25. Robert Jungk, *Brighter Than a Thousand Suns: A Personal History of the Atomic Scientists* (New York: Harcourt Brace Jovanovich, 1958); Kai Bird and Martin J. Sherwin, *American Prometheus: The Triumph and Tragedy of J. Robert Oppenheimer* (New York: Knopf, 2005).
26. Correspondence from K. T. Bainbridge to Charles Yulish, March 22, 1954, Bainbridge Papers, box 31, folder "Correspondence, 1943–1948."
27. William L. Laurence, *Dawn over Zero: The Story of the Atomic Bomb*, 2nd ed. (New York: Knopf, 1947), 191.
28. Correspondence from K. T. Bainbridge to Mr. Cahn, April 30, 1960, reproduced in Reminiscences of Kenneth Tompkins Bainbridge, Columbia Center for Oral History, Columbia University, New York, 3.
29. "Army Ammunition Explosion Rocks Southwest Area," *El Paso Herald-Post*, July 16, 1945.
30. J. E. Warner, press memorandum, July 16, 1945, Office of Censorship Files, box 481, file "Atom Smashing May 1945."
31. Michael S. Sweeney, *Secrets of Victory: The Office of Censorship and the American Press and Radio in World War II* (Chapel Hill: University of North Carolina Press, 2001);

"Alamogordo Base Explosives Blast Jolts Wide Area," *Albuquerque Journal*, July 17, 1945; "Airbase Blast Stirs Report of Earthquake," *Tucson Daily Citizen*, July 16, 1945.

32. A History of the Office of Censorship, vol. 2: Press and Broadcasting Division, 1945, Office of Censorship Files, entry 4, box 4, 184.

33. Lamont, *Day of Trinity*.

34. Laurence interview with Bruns; Laurence, *Men and Atoms*, 119.

35. Laurence, *Men and Atoms*, 122.

36. US Department of Energy, "Evaluations of Trinity," https://www.osti.gov/opennet/manhattan-project-history/Events/1945/trinity_evaluations.htm (accessed December 23, 2021).

37. William L. Laurence, *Dawn over Zero: The Story of the Atomic Bomb* (New York: Knopf, 1946), 128.

38. Meyer Berger, *The Story of the New York Times, 1851–1951* (New York: Simon & Schuster, 1951), 516.

39. Laurence interview with Bruns, 327.

40. Laurence interview with Bruns.

41. Laurence, *Men and Atoms*, 146.

42. Laurence, *Dawn over Zero*.

43. Joseph Laurance Marx, *Seven Hours to Zero* (New York: Putnam, 1967), 132.

44. Laurence interview with Bruns.

45. Message from Farrell to O'Leary, Manhattan District Records, box 20, Tinian Files, envelope G.

46. Message APCOM 5313, August 7, 1945, Manhattan District Records, box 19, file "These documents from folder marked—Tinian—Operational info Farrell Group."

47. Deirdre Carmody, "Hiroshima A—Bomb Log Nets $37,000," *New York Times*, November 24, 1971.

48. Laurence interview with Bruns, 331.

49. Vincent A. Sellers, oral history interview with Lawrence M. Langer (Bloomington, IN: Indiana University Oral History Research Center, 1982), 21, Indiana University oral histories accession #82-61-1.

50. *The Rising Sun Sets: The Complete Story of the Bombing of Nagasaki*, ed. Jerome Beser and Jack Spangler (Bloomington, IN: Authorhouse, 2007), 115.

51. Gordon Thomas and Max Morgan Witts, *Enola Gay* (New York: Stein & Day, 1977).

52. Laurence interview with Bruns, 337.

53. Sellers interview with Langer, 24.

54. Fletcher Knebel and Charles W. Bailey, *No High Ground* (New York: Harper, 1960).

55. "When Uranium Was 'Tuballoy,'" *Nucleonics* 12, no. 10 (1954): 20–21.

56. Vincent C. Jones, *Manhattan: The Army and the Atomic Bomb* (Washington, DC: Government Printing Office, 1985).

57. Laurence interview with Bruns.

58. Laurence, *Dawn over Zero*, 223.

59. Turner Catledge, *My Life and the Times* (New York: Harper & Row, 1971), 174.

60. Catledge, *My Life and the Times*, 175.

61. Berger, *Story of the New York Times*, 520.

62. Berger, *Story of the New York Times*, 521.

63. Theodore Frederic Koop, *Weapon of Silence* (Chicago: University of Chicago Press, 1946).

64. Koop, *Weapon of Silence*, 281.

65. Chiles Coleman, "The Day They Announced the Atomic Bomb," *Wisconsin State Journal* (Madison), July 26, 1970.

66. War Department, Information Approved for Publication on Atomic Bombs, Truman files, Part 3: Subject File, reel 41, Atomic Bomb—Press Release (folder 1).

67. "Science and the Bomb," *New York Times,* August 7, 1945, 22.

68. "War Department Called Times Reporter to Explain Bomb's Intricacies to Pubic," *New York Times,* August 7, 1945.

69. Berger, *Story of the New York Times.*

70. Laurence interview with Bruns, 394.

71. "First Man-Made Atomic Explosion in New Mexico Desert Heralds Man's Arrival at 'Atomic Age,'" *Oak Ridge (TN) Journal,* August 9, 1945.

72. Louis Liebovich, *The Press and the Origins of the Cold War, 1944–1947* (New York: Praeger, 1988), 86.

73. L. R. Groves, *Now It Can Be Told* (New York: Harper & Row, 1962), 331.

74. L. R. Groves, Reference is made to Captain Wallace's letter of March 19, 1948, Groves Papers, entry 4, box 2, folder "Atomic Energy Data," 5, 8–9.

75. Kenneth D. Nichols, *The Road to Trinity* (New York: Morrow, 1987), 202.

76. Laurence interview with Bruns.

77. Apcom 5397, Manhattan District Records, box 17, file "Gen. Farrell."

78. Ann Jacob Stocker, "A Study of the Life and Work of William L. Laurence as Related to His Contributions in Science Journalism" (master's thesis, Syracuse University, 1966), 81.

79. Laurence interview with Bruns, 357.

80. Laurence interview with Bruns, 359.

81. Clair Stebbins and Harry Franken, interviews with Paul W. Tibbets, Tibbets Papers, Ohio Historical Society, folder 2, 159.

82. Laurence interview with Bruns.

83. Robert Krauss and Amelia Krauss, eds., *The 509th Remembered: A History of the 509th Composite Group as Told by the Veterans Themselves, 509th Anniversary Reunion, Wichita, Kansas October 7–10, 2004,* rev. ed. (Buchanan, MI: 509th Press, 2010).

84. Stocker, "Study of the Life and Work," 83.

85. Laurence interview with Bruns, 371.

86. Charles W. Sweeney, James A. Antonucci, and Marion K. Antonucci, *War's End: An Eyewitness Account of America's Last Atomic Mission* (New York: Avon Books, 1997).

87. Laurence interview with Bruns, 370.

88. Laurence interview with Bruns, 379–80.

89. William L. Laurence, Message APCOM 5575, August 9, 1945, Manhattan District Records, box 17, file "Tinian Files - Index to Envelopes A to I."

90. Krauss and Krauss, "The 509th Remembered," 212.

91. Paul Ham, *Hiroshima Nagasaki* (New York: Thomas Dunne Books, 2011).

92. William L. Laurence, "Atomic Bombing of Nagasaki Told by Flight Member," *New York Times,* September 9, 1945.

93. Laurence interview with Bruns, 392.

94. Sweeney et al., *War's End,* 230.

95. Laurence interview with Bruns.

96. "Science of Reporting," *Time,* December 27, 1963, 32–33.

97. Daily Diary, 9 August 1945, Spaatz Papers, box I:21, file "August 1945."

98. "Atomic Bomb 'Short Snorter' Is Lt. Del Genio's Top Relic," *Knoxville (TN) News-Sentinel,* October 25, 1945.

99. William J. Wilcox Jr., "From Us in Oak Ridge to Tojo: The Story of Lt. Nicholas Del Genio's Autographed Dollar Bill," American Museum of Science and Energy Foundation, June 15, 2000, http://paperzz.com/doc/8177552/how-the-last-piece-of-little-boy-got-to-tinian.

100. Samuel Martin Lebovic, "Fighting for Free Information: American Democracy and the Problem of Press Freedom in a Totalitarian Age, 1920–1950" (PhD diss., University of Chicago, 2011), 351.

101. Sweeney, *Secrets of Victory.*

102. Arthur Gelb, *City Room* (New York: Putnam, 2003), 102.

103. Wilfred G. Burchett, *Shadows of Hiroshima* (London: Verso, 1983), 16.

104. Peter B. Hales, "The Atomic Sublime," *American Studies* 32, no. 1 (1991): 12–13.

105. Stewart L. Udall, *The Myths of August: A Personal Exploration of Our Tragic Cold War Affair with the Atom* (New York: Pantheon Books, 1994), 25.

106. Alex Wellerstein, *Restricted Data: The History of Nuclear Secrecy in the United States* (Chicago: University of Chicago Press, 2021), 112; Cornelia Dean, *Am I Making Myself Clear? A Scientist's Guide to Talking to the Public* (Cambridge, MA: Harvard University Press, 2009), 43.

107. Amy Goodman and Juan Gonzalez, "The Hiroshima Cover-Up," *Baltimore Sun,* August 5, 2005.

108. "Duty of Journalists in a Time of War," *Virginian-Pilot* (Norfolk), June 23, 2005.

109. Jack Shafer, "Blame it on the *New York Times,*" *Slate,* April 14, 2005.

110. Laurence interview with Bruns, 396–97.

111. Message APCOM 400, August 18, 1945, Manhattan District Records, box 17, Tinian files, envelope G.

112. Message WAR 53682, August 23, 1945, Manhattan District Records, box 17, Tinian files, envelope G.

113. Message APCOM 5867, August 24, 1945, Manhattan District Records, box 17, Tinian files, envelope G.

114. INS Records, NAI Number: 4490795.

6. Aftermath

1. Arthur Gelb, *City Room* (New York: Putnam, 2003), 104.

2. William Leonard Laurence, interview by Scott Bruns, Columbia Center for Oral History, Columbia University, New York, 400.

3. Message APCOM 5575, August 9, 1945, Manhattan District Records, box 17, file "Tinian Files—Index to Envelopes A to I"; William L. Laurence, "Atomic Bombing of Nagasaki Told by Flight Member," *New York Times,* September 9, 1945.

4. Eye Witness Account, Atomic Bomb Mission over Nagasaki, Truman Files, Part 3: Subject File, reel 41, Atomic Bomb—Press Release (folder 1).

5. Laurence, "Atomic Bombing of Nagasaki."

6. "Pillar of Purple Fire Shoots 10,000 Feet into Skies over Nagasaki When Atom Bomb Is Dropped," *Montana Standard* (Butte), September 9, 1945; "Atomizing of Nagasaki Told," *Oakland Tribune,* September 9, 1945; "10,000-Foot Pillar of Fire Is Seen Following Atomic Bomb Blast," *Brownsville (TX) Herald,* September 9, 1945; "Fire Rises 10,000 Feet as A-Bomb Hits Nagasaki," *Knoxville (TN) Journal,* September 9, 1945.

7. William L. Laurence, "Eyewitness Tells First Story of A-Bomb Blast," *Los Angeles Times,* September 9, 1945.

8. "Atom Bomb a 'Man-made Meteor,'" *Sunday Times,* September 9, 1945; "He Saw Atom Bomb Burst," *Sunday Post* (Glasgow, Scotland), September 9, 1945.

9. " 'Eyewitness' Description of Atomic Bombing," *Northern Miner* (Charters Towers, Queensland, Australia), September 10, 1945.

10. City of New York, Office of the Mayor, text of Mayor F. H. LaGuardia's Sunday Broadcast to the People of New York from His Office in City Hall, Sunday, September 9, 1945, LaGuardia Papers, box #26C2, folder #31, 4.

11. W. H. Lawrence, "Visit to Hiroshima Proves It World's Most-Damaged City," *New York Times,* September 5, 1945.

12. W. H. Lawrence, "No Radioactivity in Hiroshima Ruin," *New York Times,* September 13, 1945.

13. "53,000 Dead at Hiroshima," *Times* (London), September 6, 1945.

14. Wilfred G. Burchett, *Shadows of Hiroshima* (London: Verso, 1983), 17.

15. William L. Laurence, "Nagasaki Was the Climax of the New Mexico Test," *Life,* September 24, 1945, 30.

16. William L. Laurence, "U.S. Atom Bomb Site Belies Tokyo Tales," *New York Times,* September 12, 1945.

17. Correspondence from L. R. Groves to Arthur H. Sulzberger, September 12, 1945, Sulzberger Papers, box 109, folder 2.

18. Correspondence from Arthur Hays Sulzberger to Major General L. R. Groves, September 27, 1945, Sulzberger Papers, box 109, folder 2.

19. William L. Laurence, "Lightning Blew Up Dummy Atom Bomb," *New York Times,* September 27, 1945; William L. Laurence, "Atom Bomb Based on Einstein Theory," *New York Times,* September 28, 1945; William L. Laurence, "Atomic Factories Incredible Sight," *New York Times,* September 29, 1945; William L. Laurence, "Engineering Vision in Atomic Project," *New York Times,* October 1, 1945; William L. Laurence, "Gases Explain Size of Atomic Plants," *New York Times,* October 3, 1945; William L. Laurence, "Scientists 'Create' in Atomic Project," *New York Times,* October 4, 1945; William L. Laurence, "Element 94 Key to Atomic Puzzle," *New York Times,* October 5, 1945; William L. Laurence, "Plutonium Lifted by New Chemistry," *New York Times,* October 8, 1945; William L. Laurence, "Atomic Key to Life Is Feasible Now," *New York Times,* October 9, 1945.

20. Correspondence from Edwin James to L. J. Groves, September 14, 1945, Groves Papers, entry 4, box 1, folder "MED—Correspondence (Misc)."

21. "Atom Bomb Forum," *New York Times,* November 20, 1945.

22. Laurence interview with Bruns, 404.

23. Robert Simpson, "The Infinitesimal and the Infinite," *New Yorker,* August 18, 1945, 26.

24. Jo Ranson and Richard Pack, *Opportunities in Radio* (New York: Vocational Guidance Manuals, 1946), 27.

25. "Victory Loan Smashes over the Top," *Minute Man,* December 15, 1945, 4–9; "Book and Author Rallies," *Minute Man,* November 15, 1945, 24.

26. "Bill Mauldin, Three Authors in Cleveland as Part of Whirlwind Bond Selling Tour," *Cleveland Press,* November 2, 1945.

27. "Atom in Spotlight at War Bond Show," *Gary (IN) Post-Tribune,* November 5, 1945, 5.

28. Mutual's *Meet the Press*, Friday, November 16, 1945, 10:00–10:30 p.m., Spivak Papers, box 199.

29. "Million Lives Saved by Atomic Bomb Says N.Y. Writer," *Dunkirk (NY) Evening Observer,* November 26, 1945.

30. "Usable Atom Power Is Predicted Near," *Arizona Republic* (Phoenix), December 7, 1945.

31. William L. Laurence, "12:01 World Time," *Survey Graphic,* January 1946.

32. Frank Sullivan, "The Cliché Expert Testifies on the Atom," *New Yorker,* November 17, 1945, 27, ellipses in original.

33. Correspondence from Alfred A. Knopf to Lee Roy Whitman, August 22, 1945, Knopf Records, box 8, folder 8.

34. Correspondence from LeRoy Whitman to Alfred A. Knopf, August 28, 1945; correspondence from Alfred A. Knopf to William L. Laurence, August 23, 1945; Knopf Records, box 8, folder 8.

35. Correspondence to Alan Collins, September 6, 1945, Knopf Records, box 8, folder 8.

36. Correspondence from William A. Koshland to Naomi Burton, August 22, 1946, Knopf Records, box 8, folder 8.

37. Correspondence from Bernard Smith to Alan Collins, March 19, 1946, Knopf Records, box 8, folder 8.

38. William L. Laurence, "The Atom Today and Tomorrow," in *Information Please Almanac 1947*, ed. John Kieran (New York: Doubleday, 1947), 12.

39. Bruce Rae, "Proving That Facts Can Be Fun," *New York Times,* January 26, 1947.

40. Elizabeth S. Kingsley, "Double-Crostic," *New York Times,* August 3, 1947; "Solution to Last Week's Double-Crostic Puzzle," *New York Times,* August 10, 1947.

41. William L. Laurence, "Tomorrow You May Be Younger," *Ladies' Home Journal,* December 1945, 22–23.

42. William L. Laurence, "Tomorrow You May Be Younger," *Reader's Digest,* February 1946.

43. "You Don't Have to Get Old Says Professor; Serum, ACS, Thwarts Old-Age Ills," *Indiana (PA) Evening Gazette,* January 28, 1946; William L. Laurence, "New Weapon against Death," in *Science Year Book of 1947*, ed. J. D. Ratcliffe (Garden City, NY: Doubleday, 1947), 58–68.

44. Thomas S. Gardner, "The Prolongation of Life," *Journal of Gerontology* 1, no. 4 (1946): 228–29; "Current Comment," *Journal of Gerontology* 1, no. 2, part 1 (April 1, 1946): 252–53.

45. Waldemar Kaempffert, "Science in Review," *New York Times,* May 5, 1946.

46. Drew Middleton, "Soviet Biologist Sees 150-Year Life If His New Serum Is Used Properly," *New York Times,* June 7, 1946.

47. "Bogomolets' Serum," *New York Times,* June 8, 1946.

48. Waldemar Kaempffert, "How to Grow Old, and Like It," *New York Times,* June 16, 1946.

49. Leslie Brent, *A History of Transplantation Immunology* (San Diego: Academic Press, 1997), 250–51.

50. David Stipp, *The Youth Pill: Scientists at the Brink of an Anti-Aging Revolution* (New York: Current, 2010).

51. Laurence, "Plutonium Lifted by New Chemistry."

52. Paul Friggens, Report of the Journalism Faculty Committee on Reporting Nominations for 1946 Pulitzer Prizes for Work done in 1945, April 1, 1946, Pulitzer archives.

53. Friggens, Report of the Journalism Faculty Committee on Reporting Nominations for 1946 Pulitzer Prizes.

54. "Pulitzer Prizes Awarded; 'State of Union' the Play," *New York Times,* May 7, 1946.

55. *Boston University Commencement Exercises, Nineteen Hundred Forty-Six* (Boston: Boston University, 1946).

56. Correspondence from L. R. Groves to Edwin L. James, May 10, 1946, and correspondence from L. R. Groves to William L. Laurence, May 10, 1946, Groves Papers, entry 4, box 1, folder "MED—Correspondence (Misc)."

57. Correspondence from William L. Laurence to General Leslie R. Groves, May 14, 1946, Groves Papers, entry 4, box 1, folder "MED—Correspondence (Misc)."

58. Oral history of Turner Catledge, Times oral histories, box 2, folder 2, 201.

7. Atomic Plagiarism in the South Pacific

1. Correspondence from Walter Sullivan to Florence Laurence, March 31, 1977, Sullivan Papers, box 48, folder 3; David M. Rubin and Ann Marie Cunningham, "War, Peace & the News Media: Proceedings, March 18 and 19, 1983" (New York University, 1983), 214.

2. Bob Considine, "Atoms Aweigh," in *Deadline Delayed*, ed. Overseas Press Club of America (New York: E. P. Dutton, 1947), 33–44.

3. Commander Joint Task Force One, Orders to William L. Laurence, Defense Nuclear Agency Records, entry 48C, box 200, folder "Lab-May."

4. Howard W. Blakeslee, "Activity at 'Crossroads,'" *AP World*, Autumn 1946, 12.

5. Robert D. Potter, History of NASW, NASW Papers, box 4, folder 27, 44.

6. Howard W. Blakeslee, "Expert Sees No Value to Science or Military in Bikini Bomb Test," *Titusville (PA) Herald*, May 30, 1946; Lee A. DuBridge, "What About the Bikini Tests?" *Bulletin of the Atomic Scientists* 1, no. 11 (May 15, 1946): 7–16.

7. L. A. DuBridge, "Atomic Test Queried," *New York Times*, May 5, 1946.

8. William L. Laurence, "Stars' Secrets May Unfold to Myriad Gauges at Bikini," *New York Times Magazine*, April 24, 1946.

9. Anatol James Shneiderov, "Nuclear Fission Bomb as Initiator of Earthquakes," *Science* 103, no. 2683 (May 31, 1946): 675.

10. Howard W. Blakeslee, "Warning Given Atom Bomb Fleet; No Survivors for Report Feared," *Charleston (WV) Daily Mail*, June 11, 1946.

11. Howard W. Blakeslee, "All A-Bomb Personnel May Die, Scientist Says," *San Francisco Examiner*, June 11, 1946; "The Bikini Tests May Split Earth, Scientist Predicts," *San Francisco Chronicle*, June 11, 1946.

12. Jim G. Lucas, "Long Faces," *Washington Daily News*, June 21, 1946; "Bikini: Breath-Holding before a Blast—Could It Split the Earth?" *Newsweek*, July 1, 1946.

13. "Crews Reassured About Atom Bomb," *Milwaukee Journal*, June 13, 1946.

14. Potter, History of NASW, 42.

15. Jonathan M. Weisgall, *Operation Crossroads: The Atomic Tests at Bikini Atoll* (Annapolis, MD: Naval Institute Press, 1994), 151.

16. John G. Harris, "Globe Man Bikini-Bound," *Boston Globe*, June 18, 1946.

17. Robert U. Brown, "Shop Talk at Thirty," *Editor and Publisher*, June 22, 1946, 64.

18. Vincent Tubbs, "Tubbs a Bit Disappointed by Atom Bomb Test Results," *Afro-American*, July 13, 1946.

19. Press Filing #847, June 25, 1946, Defense Nuclear Agency Records, entry 47, box 87, file "25 June 1946, 800–869."

20. H. C. Cleavinger, "Washington Is Tabbed Atom Capital of World," *Spokane Daily Chronicle*, June 22, 1946.

21. W. H. Lawrence, "Baruch Urges Pact," *New York Times*, June 15, 1946.

22. Weisgall, *Operation Crossroads*, 151.

23. Louis Spilman, "'Operation Crossroads': Eye-Witness Story of Atom Bomb Tests at Bikini Island," *Free Lance-Star* (Fredericksburg, VA), June 27, 1946.

24. E. W. MacAlpine, "U.S. Atomic Plan 'Bombshell' for Bikini Men," *Argus* (Melbourne, Victoria), June 17, 1946.

25. "Laurence Discusses Atomic Energy Here," *Trinity Tripod*, December 14, 1946.

26. Beverly Benson, "Laurence Calls on Americans to Back Baruch Control Plan," *Daily Iowan* (Iowa City), November 22, 1946; "Writer Blasts Russ View of Atomic Bomb," *Salt Lake Tribune,* November 10, 1946.

27. "Atom Control Plea Made by Laurence," *New York Times,* September 19, 1946.

28. Walter W. Ruch, "Motor Luncheon Is Held in Detroit," *New York Times,* November 15, 1946.

29. "Honolulu Plays Host to Visiting A-Bomb Personnel," *Honolulu Star-Bulletin,* June 19, 1946.

30. S. Burton Heath, "Gladys—Atom Experts Stumped by Her Cravings," *Indiana Evening Gazette,* July 29, 1946.

31. Vincent Tubbs, "Ace Newsman on A-Bomb Test Give Odds to Joe Louis," *Afro-American,* June 22, 1946; James P. Dawson, "Louis Stops Conn in Eighth Round and Retains Title," *New York Times,* June 20, 1946.

32. Lucas, "Long Faces."

33. Philip W. Porter, "Porter and Other Writers Briefed on Bikini Blasts," *Plain Dealer* (Cleveland, OH), June 22, 1946.

34. Hector Stewart, Press Filing #246, June 21, 1946, Defense Nuclear Agency Records, entry 47, box 86A, file "21 June 1946, Number 170–257 [Part 1 of 2]."

35. Associated Press article, June 23, 1946, Downs Papers, box 2, folder 35.

36. Bob Considine, Press Filing #389, June 23, 1946, Defense Nuclear Agency Records, entry 47, box 87, file "23 June 1946, Nos. 368–481 [Part 2 of 2]."

37. William L. Laurence, "Bikini Gears Mesh in Vast Test Plan," *New York Times,* June 26, 1946.

38. Frank H. Bartholomew, *Bart: Memoirs of Frank H. Bartholomew* (Sonoma, CA: Vine Brook Press, 1983), 131.

39. William L. Laurence, Press Filings #1661–1663, June 29, 1946, Defense Nuclear Agency Records, entry 47, box 89, file "29 June 1946, Nos. 1637 to 1703."

40. William L. Laurence, "Ships in Place for Bomb," *New York Times,* June 30, 1946.

41. Edward F. Jones, Press Filing #1484, Defense Nuclear Agency Records, entry 47, box 94B, file "7–10 July 1946, Nos. 1473 to 1535."

42. Bob Considine, *It's All News to Me: A Reporter's Deposition* (New York: Meredith Press, 1967), 202.

43. Telegram from William L. Laurence to Mr. James, June 26, 1946, Catledge Papers, folder "Laurence William L., 1946."

44. Correspondence from Turner Catledge to Mr. Henderson, June 26, 1946, and telegram from Turner Catledge to William L. Laurence, June 26, 1946, both in Catledge Papers, folder "Laurence William L., 1946."

45. Robert Littell, "The Voice of the Apple," *Harper's,* September 1946, 225.

46. Tubbs, "Tubbs a Bit Disappointed," 17.

47. Telegram from Lawrence E. Davies to Turner Catledge, Catledge Papers, folder "Laurence William L., 1946."

48. William L. Laurence, "Fiery 'Super Volcano' Awes Observer of 3 Atom Tests," *New York Times,* July 1, 1946.

49. Laurence, "Fiery 'Super Volcano.' "

50. "Nevada Survives," *New York Times,* July 1, 1946.

51. Cable from Wogonie to Frank Bartholomew, Bill Tyree, Joe Myler, and Robert Bennyhoff, July 2, 1946, Defense Nuclear Agency Records, entry 47A, box 68.

52. Telegram from Turner Catledge to William L. Laurence, July 1, 1946, Catledge Papers, folder "Laurence William L., 1946."

53. Cable from Larry Davies to William L. Laurence, July 2, 1946, Defense Nuclear Agency Records, entry 47A, box 68.

54. William E. Beard, "Banner Editor Likens Atomic Glow to That of Glaring Sun," *Nashville Banner,* July 1, 1946; William E. Beard, Press Filing #344, July 1, 1946, Defense Nuclear Agency Records, entry 47A, box 92, file "1 July 1946, Nos. 301 to 390, Part 2 of 2."

55. Keyes Beech, "Many Observers Are Lukewarm in Reaction to Big Show on Bikini," *Honolulu Star-Bulletin,* July 1, 1946.

56. US Department of Energy, *United States Nuclear Tests, July 1945 through September 1992* (Las Vegas: US Department of Energy Nevada Operations Office, 2015), https://www.nnss.gov/docs/docs_LibraryPublications/DOE_NV-209_Rev16.pdf (accessed May 29, 2020).

57. Littell, "Voice of the Apple," 226.

58. William F. Tyree, "Bomb Fails to Wipe Out Bikini Atoll," *New York Herald Tribune,* July 1, 1946.

59. Frank H. Bartholomew, Press Filing #1115, July 4, 1946, Defense Nuclear Agency Records, entry 47A, box 94, file "4 July 1946, Nos. 1109 to 1289, Part 3 of 3"; Frank H. Bartholomew, "Bomb Couldn't Equal Buildup," *Tucson Daily Citizen,* July 8, 1946.

60. Clark Lee, *One Last Look Around* (New York: Duell, Sloan & Pearce, 1947), 294.

61. William L. Laurence, *Dawn over Zero: The Story of the Atomic Bomb,* 2nd ed. (New York: A. A. Knopf, 1947), 275.

62. "Atom Research Men Satisfied," *Telegraph-Herald* (Dubuque, IA), July 1, 1946; "A-Bomb No Failure, Scientist Says," *El Paso (TX) Herald-Post,* July 2, 1946.

63. Weisgall, *Operation Crossroads.*

64. "Results at Bikini," *New York Times,* July 2, 1946.

65. William L. Laurence, "Bomb at Bikini Exploded Too Low, Reducing Effect, Experts Indicate," *New York Times,* July 5, 1946.

66. Defense Nuclear Agency, *Operation Crossroads 1946* (Washington, D.C.: Defense Nuclear Agency, 1984).

67. Telegram from Edwin L. James to William Laurence, c/o JTF-1 Public Information, July 3, 1946, Defense Nuclear Agency Records, entry 47, box 69, binder "July 3, 1946."

68. Message #0046 to CJTF-1, July 4, 1946, Defense Nuclear Agency Records, entry 47, box 69, binder "July 4, 1946."

69. Telegram from Shephard Stone to William Laurence, c/o JTF-1 Public Information, July 4, 1946, Defense Nuclear Agency Records, entry 47, box 69, binder "July 4, 1946."

70. William L. Laurence, Press Filing #1418, July 5, 1946, Defense Nuclear Agency Records, entry 47A, box 92, file "5 July 1946, Nos. 1289 to 1442, Part 1 of 2."

71. Telegram from Shephard Stone to William Laurence, c/o JTF-1 Public Information, July 6, 1946, Defense Nuclear Agency Records, entry 47, box 70, binder "July 6, 1946."

72. William L. Laurence, Press Filing #1419, July 5, 1946, Defense Nuclear Agency Records, entry 47, box 94, file "5 July 1946, Nos. 1289 to 1442, Part 1 of 2."

73. Telegram from Edwin L. James to William Laurence, c/o USS Appalachian, July 6, 1946, Defense Nuclear Agency Records, entry 47, box 70, binder "July 6, 1946."

74. William L. Laurence, Press Filing #664, July 2, 1946, Defense Nuclear Agency Records, entry 47, box 94, file "2 July 1946, Nos. 558 to 757, Part 2 of 3."

75. Jones, Press Filing #1484.

76. "Bikini Rehearsal Makes 'Loud Pop,' " *New York Times,* July 19, 1946.

77. JTF-1 Public Info at Kwajalein, Message to CJTF-1, July 10, 1946, Defense Nuclear Agency Records, entry 47, box 71, binder "FOX 100001 to 101153 July 10, 1946."

78. William L. Laurence, "Bikini's King Gets Truman's Thanks," *New York Times,* July 17, 1946.

79. Kenneth McArdle, " 'Baker Day' Story Reporter's Dream," *Editor and Publisher,* August 3, 1946, 64.

80. McArdle, " 'Baker Day.' "

81. William L. Laurence, "Blast Biggest Yet," *New York Times,* July 25, 1946.

82. Hanson W. Baldwin, "Lost Chance in War Set Bikini Pattern," *New York Times,* July 25, 1946.

83. Laurence, "Blast Biggest Yet."

84. Baldwin, "Lost Chance in War Set Bikini Pattern."

85. Keyes Beech, "Explosion Most Spectacular Yet Seen by Man," *Honolulu Star-Bulletin,* July 25, 1946.

86. Howard W. Blakeslee, Telegram #1514, Defense Nuclear Agency Records, entry 47, box 47A, binder "7–10 July 1946, Nos. 1453–1535."

87. Hanson W. Baldwin, "Lessons Learned in Bikini Tests," *New York Times,* August 1, 1946.

88. William L. Laurence, "Bikini 'Dud' Decried for Lifting Fears," *New York Times,* August 4, 1946.

89. Laurence, "Bikini 'Dud.' "

90. Baldwin, "Lessons Learned in Bikini Tests."

91. Hanson W. Baldwin, "Atomic Rays Deal Death Stealthily; Bikini Has Uncovered No Defenses," *New York Times,* August 2, 1946.

92. Bartholomew, *Bart,* 138–39; William L. Laurence, *Men and Atoms: The Discovery, the Uses, and the Future of Atomic Energy* (New York: Simon & Schuster, 1959), 209. Laurence wrote that the trees were called the "William L. Laurence Memorial Grove," even though he was alive when the name was devised. Considine wrote that the grove was named "Laurence Grove" (*Atoms Aweigh,* 33–44).

93. Laurence, *Men and Atoms,* 209.

8. Reporter Grade 8

1. *The Mary Margaret McBride Program,* December 2, 1947, Recorded Sound Research Center, shelf no. RWC 6754 A1.

2. William L. Laurence, "The Atom in the News," Sulzberger Papers, box 109, folder 2, 2.

3 *University of Missouri Bulletin: 1947 Missouri Honor Award,* University of Missouri Archives, Collection No. C:11/17/1.

4. "More than Explosive Destruction [Editorial]," *Joplin (MO) Globe,* February 18, 1949.

5. "Atomic Report [Review]," *Variety,* July 20, 1949, 32.

6. *Eternal Light,* August 21, 1949, Recorded Sound Research Center, shelf no. RGB 3337.

7. "A Valentine to the Following: [Advertisement]," *New York Times,* February 14, 1947.

8. "Times Writer, Wife Nabbed, Ousted from British Zone," *Buffalo (NY) Courier-Express,* July 25, 1947.

9. Alice Hughes, "Atom Bomb Becomes an Icing on a Cake—for Special Event," *Buffalo (NY) Courier-Express,* March 16, 1948.

10. William L. Laurence, "World Scientists Hail Einstein at 70," *New York Times,* March 13, 1949.

11. William L. Laurence, "New Einstein Theory Gives a Master Key to Universe," *New York Times*, December 27, 1949.

12. Peter Middleton, *Physics Envy: American Poetry and Science in the Cold War and After* (Chicago: University of Chicago Press, 2015), 102.

13. Herbert B. Nichols, "Einstein Links Atom to Cosmos," *Christian Science Monitor*, December 27, 1949; "New Discovery by Einstein May Solve Gravitation Riddle," *Washington Post*, December 27, 1949.

14. Chuck Hansen, *U.S. Nuclear Weapons: The Secret History* (Arlington, TX: Aerofax, 1988).

15. "Tests of Improved Atomic Bombs Are Reported Successful," *Boston Globe*, May 19, 1948; "Only Americans at Atomic Tests," *New York Times*, May 19, 1948; Anthony Leviero, "Truman Declares Russia Forces U.S. into Bomb Secrecy," *New York Times*, July 25, 1948; "No Bombs Used in Atoll Atom Blasts," *Syracuse (NY) Herald Journal*, May 19, 1948; Beverly Deepe Keever, *News Zero: The New York Times and the Bomb* (Monroe, ME: Common Courage Press, 2004); William L. Laurence, "A Swift Succession of Significant Events Occurs in the Field of Atomic Energy," *New York Times*, December 7, 1947.

16. "U.S. Atom Secrets Held Still Secure," *New York Times*, March 10, 1948.

17. Robert S. Norris and Hans M. Kristensen, "Global Nuclear Weapons Inventories, 1945–2010," *Bulletin of the Atomic Scientists* 66, no. 4 (July/August 2010): 77–83.

18. Norris and Kristensen, "Global Nuclear Weapons Inventories."

19. Mose [*sic*] Salisbury, letter to William L. Laurence, July 29, 1948, AEC archives, entry 1, box 7, folder "Information Division—Memoranda—1948."

20. William L. Laurence, "How Soon Will Russia Have the A-Bomb?," *Saturday Evening Post*, November 6, 1948, 23.

21. Richard Rhodes, *Dark Sun: The Making of the Hydrogen Bomb* (New York: Simon & Schuster, 1995).

22. William L. Laurence, "Soviet Achievement Ahead of Predictions by 3 Years," *New York Times*, September 24, 1949.

23. Samuel Nelson Drew, "Moving Democracy to Action: Agenda Setting and Consensus Building in Developing Responses to Perceived Soviet Threats" (PhD diss., University of Virginia, 1986).

24. S. H. Cross, letter to A. C. Hanford, March 29, 1937, and A. C. Hanford to S. H. Cross, April 2, 1937, both in Laurence Harvard file.

25. Ralph Lowell, untitled diary transcript, Lowell Papers, box 12, folder "Diary Sept.-Oct. 1947," 197.

26. Ralph Lowell, letter to William Bender, October 15, 1947, and W. J. Bender to Ralph Lowell, October 17, 1947, both in Laurence Harvard file.

27. William L. Laurence, letter to Alfred Chester Hanford, January 6, 1948, Laurence Harvard file.

28. Unsigned memorandum, January 9, 1948, Laurence Harvard file.

29. W. J. Bender, letter to William L. Laurence, January 13, 1948, Laurence Harvard file.

30. William L. Laurence, letter to W. J. Bender, January 28, 1948, Laurence Harvard file.

31. W. J. Bender, letter to William L. Laurence, February 2, 1948, Laurence Harvard file.

32. William L. Laurence, letter to W. J. Bender, March 10, 1948, Laurence Harvard file.

33. W. J. Bender, letter to Paul H. Buck, March 29, 1948, Laurence Harvard file.

34. W. J. Bender, letter to William L. Laurence, March 23, 1948, Laurence Harvard file.

35. W. J. Bender, letter to William L. Laurence, May 8, 1948, Laurence Harvard file.

36. William L. Laurence, letter to W. J. Bender, May 28, 1948, Laurence Harvard file.

37. Laurence to Bender, May 28, 1948.

38. Robert Simpson, "William the Nuclear," *Times Talk,* June 1955, 4, Sulzberger Papers, box 11, folder 13.

39. William Berkowitz, ed., *Let Us Reason Together* (New York: Crown, 1970), 11–25.

40. William Leonard Laurence, interview by Scott Bruns, Columbia Center for Oral History, Columbia University, New York.

41. "Exploring the Field of Atomic Energy," *Bostonia,* March 1949, 21.

42. Laurence, letter to Edwin L. James, February 18, 1949, Sulzberger Papers, box 42, folder 1.

43. Laurence to James, February 18, 1949, 3.

44. Laurence to James, February 18, 1949, 2.

45. Laurence to James, February 18, 1949, 4–5.

46. United States Atomic Energy Commission, Staff Conference Minutes, Public and Technical Information Service, no. 20, April 20, 1949; United States Atomic Energy Commission, Staff Conference Minutes, Public and Technical Information Service, no. 23, April 29, 1949.

47. "Employment Record, Laurence, William L.," Catledge Papers, series IIC, box 5, folder "Laurence William L., 1952–1968."

48. Edwin L. James, letter to Mr. Sulzberger, June 21, 1949, Sulzberger Papers, Manuscripts and Archives Division, New York Public Library, box 42, folder 1.

49. William L. Laurence, "Atomic Materials Seen British Quest," *New York Times,* July 20, 1949.

50. Laurence, letter to Mr. Sulzberger, July 20, 1949, Sulzberger Papers, box 42, folder 1.

51. Edwin L. James, letter to Mr. Sulzberger, July 20, 1949, Sulzberger Papers, box 42, folder 1.

52. Edwin L. James, letter to Mr. Sulzberger, July 22, 1949, Sulzberger Papers, box 42, folder 1, quotation on 2.

53. Arthur Hays Sulzberger, "Memorandum for Mr. Laurence," July 25, 1949, Sulzberger Papers, box 42, folder 1.

9. The Elixir of Life

1. Meeting of April 10, 1949, 10:00 A.M. Tentative Agreement of Policy between Mayo Clinic and Merck & Co., Inc., Concerning use of, and Release of, Information about, Compound E., Mayo Clinic archives.

2. H. F. Polley and C. H. Slocumb, "Cortisone—a Historical Discovery," 1992, Mayo Clinic archives; Thom Rooke, *The Quest for Cortisone* (East Lansing: Michigan State University Press, 1992).

3. Polley and Slocumb, "Cortisone"; H. F. Polley and C. H. Slocumb, "Behind the Scenes with Cortisone and ACTH," *Mayo Clinic Proceedings* 51, no. 8 (August 1976): 471–77.

4. Polley and Slocumb, "Cortisone."

5. Polley and Slocumb, "Cortisone"; Polley and Slocumb, "Behind the Scenes with Cortisone and ACTH."

6. Rooke, *Quest for Cortisone.*

7. Polley and Slocumb, "Cortisone."

8. Edward C. Kendall, *Cortisone* (New York: Charles Scribner's Sons, 1971).

9. Rooke, *Quest for Cortisone.*

10. William L. Laurence, "Miracle Relief from Arthritis," *Ladies' Home Journal,* August 1949.

11. Rooke, *Quest for Cortisone.*

12. Laurence, "Miracle Relief from Arthritis."
13. William L. Laurence, "Aid in Rheumatoid Arthritis Is Promised by New Hormone," *New York Times*, April 21, 1949.
14. Polley and Slocumb, "Cortisone."
15. "New Recipe Eases Pain of Arthritis," *Washington Post*, April 22, 1949; "Hormone Aid for Arthritis Victims Bared," *Chicago Daily Tribune*, April 22, 1949.
16. "Mayo Clinic Physicians Find Possible Cure for Arthritis," *Richmond (VA) Times-Dispatch*, April 21, 1949.
17. "Clinic Reveals 'Effective' Use of Compound," *Rochester (MN) Post-Bulletin*, April 21, 1949.
18. William L. Laurence, "Hormone Held Aid in Rheumatic Ills," *New York Times*, April 22, 1949; "New Hope for Arthritics," *New York Times*, April 22, 1949; Waldemar Kaempffert, "Hormone Promises Aid in Arthritis," *New York Times*, April 24, 1949.
19. N. S. Haseltine, "New Remedy Is Studied for Arthritis," *Washington Post*, May 11, 1949.
20. "Experts to Report on Rheumatic Ills," *New York Times*, May 29, 1949; Howard A. Rusk, "Medical Parley Specialists Optimistic over Arthritis," *New York Times*, May 29, 1949.
21. N. S. Haseltine, "Report Due This Week on Use of Arthritis 'Magic Elixir,'" *Washington Post*, May 29, 1949.
22. Robert C. Boardman, "Movies Reveal Aid in Arthritis by Compound E," *New York Herald-Tribune*, June 1, 1949.
23. "Mayo Doctors Tell of a Remedy for 'Worst' Arthritis," *Chicago Daily Tribune*, June 1, 1949.
24. William L. Laurence, "More Relief Found in Arthritis Cases," *New York Times*, June 1, 1949; William L. Laurence, "Arthritis Remedy May Aid Epileptics," *New York Times*, June 2, 1949; Waldemar Kaempffert, "Hormone Treatment for Arthritis Promises to Bring a Baffling Disease under Control," *New York Times*, June 5, 1949.
25. "Laurence Drops Seed in Truman's Lap," *Times Talk*, August 1949, 1.
26. Margaret B. Kreig, *Green Medicine: The Search for Plants That Heal* (Chicago: Rand McNally, 1964).
27. Correspondence from William L. Laurence to Charles G. Ross, June 21, 1949, Sulzberger Papers, box 41, folder 24, "Laurence, William L. 1951–1968."
28. "Laurence Drops Seed in Truman's Lap," 1–2.
29. William L. Laurence, Confidential Memorandum for the President, June 27, 1949, Sulzberger Papers, box 41, folder 24, "Laurence, William L. 1951–1968."
30. Laurence, Confidential Memorandum for the President.
31. Laurence, Confidential Memorandum for the President.
32. "Laurence Drops Seed in Truman's Lap."
33. "The President's Day [June 28, 1949]," Harry S. Truman Library and Museum, https://www.trumanlibrary.gov/calendar?month=6&day=28&year=5 (accessed May 30, 2020).
34. "Laurence Drops Seed in Truman's Lap."
35. Correspondence from Leonard Scheele to Mrs. Albert D. Lasker, July 19, 1949, Lasker Papers, box 141, file "Cortisone (and ACTH) 1949."
36. "Laurence Drops Seed in Truman's Lap." By contrast, Kreig says that Scheele tipped off Laurence that an unnamed federal official—presumably Ewing—was planning to announce the story without giving Laurence any notice, effectively robbing him of his exclusive. Kreig, *Green Medicine*, 34.
37. William L. Laurence, "Arthritis Remedy in Quantity Promised by African Seed," *New York Times*, August 16, 1949.

38. Bess Furman, "Group Sent to Hunt Cortisone in Africa," *New York Times,* August 17, 1949.

39. "Plant Offers New Source of Arthritis-Fighting Drug," *Cedar Rapids (IA) Gazette,* August 16, 1949.

40. Herm Sittard, "Seed of an African Plant Revealed Plentiful Source of Compound 'E,'" *Rochester (MN) Post-Bulletin,* August 16, 1949.

41. James Eckman, Memorandum: Cortisone and Story by William L. Laurence on Strophanthus, August 16, 1949, Mayo Clinic archives.

42. "African Treasure Hunt," *Newsweek,* August 29, 1949, 42.

43. "Short Cut?" *Time,* August 29, 1949, 63.

44. "U.S. Scientists in Africa Hunting Rare Plant to Help Fight Arthritis," *Los Angeles Times,* August 17, 1949.

45. "Seed of Rare Plant New Hope for Arthritics," *Chicago Daily Tribune,* August 16, 1949.

46. "Scientists Seek Drug Which Cures Arthritis," *Statesville (NC) Record and Landmark,* August 17, 1949.

47. "Cortisone Source Is Growing in City," *New York Times,* August 18, 1949.

48. William L. Laurence, "Discovery of Cortisone-Yielding Plant Adds to the Great Advances against Arthritis," *New York Times,* August 21, 1949; "The 'Elixir of Life,'" *New York Times,* August 21, 1949.

49. "Anti-Arthritic Wonder Drug Found in Yams," *Chicago Daily Tribune,* August 28, 1949; Alton L. Blakeslee, "Yam Seen as Source of New Arthritis Drug," *Los Angeles Times,* August 28, 1949.

50. "Cortisone (Cont'd)," *Time,* September 5, 1949, 49.

51. Waldemar Kaempffert, "Cortisone Made from Mexican Yams," *New York Times,* August 28, 1949; "New Arthritis Aid Is Found in Yams," *New York Times,* August 28, 1949.

52. Correspondence from Orvil E. Dryfoos to Arthur Hays Sulzberger, August 18, 1949, Sulzberger Papers, box 41, folder 24, "Laurence, William L. 1951–1968."

53. "A 'Scoop' That Aids Humanity," *New York Times,* August 27, 1949.

54. Correspondence from Arthur Hays Sulzberger to William L. Laurence, August 29, 1949, and William L. Laurence to Arthur Hays Sulzberger, September 16, 1949, both in Sulzberger Papers, box 41, folder 24, "Laurence, William L. 1951–1968."

55. "Laurence Drops Seed in Truman's Lap."

56. Correspondence from Turner Catledge to Arthur Hays Sulzberger, August 25, 1949, Sulzberger Papers, box 41, folder 24, "Laurence, William L. 1951–1968."

57. Correspondence from Arthur Hays Sulzberger to Ruth Adler, August 29, 1949, Sulzberger Papers, box 41, folder 24, "Laurence, William L. 1951–1968."

58. Correspondence from Ruth Adler to Arthur Hays Sulzberger, August 29, 1949, Sulzberger Papers, box 41, folder 24, "Laurence, William L. 1951–1968."

59. Loren Ghiglione, *Radio's Revolution: Don Hollenbeck's CBS Views the Press* (Lincoln: University of Nebraska Press, 2008).

60. "Plant Hormone Discovery Made by Reporter," *Editor and Publisher,* August 20, 1949, 4.

61. "Cortisone Patent Will Be Shared," *New York Times,* September 8, 1949; correspondence from S. Blake Yates to Helen Zikmund, September 9, 1949, and memorandum by Helen L. Zikmund, September 9, 1949, both in Lasker Papers, box 141, file "Cortisone (and ACTH) 1949."

62. William L. Laurence, "Relief for 90% of Arthritics Found in Large Doses of the Sex Hormones," *New York Times,* October 2, 1949.

63. "Cortisone-Like Compound Is Yielded by Soybeans," *New York Times,* October 1, 1949; Clayton V. Sutton, "Soy Drugs Give Hope of Cheaper Relief for Arthritis Sufferers," *Wall Street Journal,* October 1, 1949.

64. "Rare Products Yield Two Hormones with High Hope of Arthritis Control," *Los Angeles Times,* August 21, 1949; Frank Carey, "Two Drugs Give New Hope in Arthritis," *Washington Post,* September 11, 1949.

65. Roy Gibbons, "Priceless New Hormone Seen as Magic Key," *Chicago Daily Tribune,* September 25, 1949.

66. William S. Barton, "Early Arthritis Relief Seen," *Los Angeles Times,* November 6, 1949.

67. Waldemar Kaempffert, "Science in Review: Production of Cortisone on a Large Scale from Western Hemisphere Plants Is Forecast," *New York Times,* November 13, 1949.

68. Department of Labor—Federal Security Agency Appropriations for 1951, pt. 2: Federal Security Agency, Public Health Service (pt. 1), 81st Cong., 2nd sess., 1950.

69. Federal Security Agency Appropriations for 1951, pt. 2.

70. "Cortisone Growth Held Years Away," *New York Times,* January 2, 1950.

71. Waldemar Kaempffert, "African Cortisone," *New York Times,* January 8, 1950.

72. "Cortisone Seekers Back from Africa," *New York Times,* June 22, 1950.

73. Waldemar Kaempffert, "African Cortisone Plants Are Here," *New York Times,* June 25, 1950.

74. Waldemar Kaempffert, "Science in Review," *New York Times,* February 26, 1950.

75. Correspondence from Philip S. Hench to Atch and Jim, March 5, 1950, Hench Papers, box 5, folder "1950, March 5—ALS & Newspaper clipping from Philip S. Hench for Atcheson Hench."

76. Correspondence from Anne Louise Davis to Arthur Hays Sulzberger, April 18, 1954, Catledge Papers, series IIA, folder "Laurence William L., 1954."

77. William L. Laurence, "New Hope for Arthritics," *Life and Health,* January 1950.

78. William L. Laurence, "You're Going to Live Longer Than You Think," *Saturday Evening Post,* April 29, 1950.

79. "Science Writers Receive Awards," *New York Times,* April 26, 1950.

80. Telegram from William L. Laurence to Edward Kendall, October 26, 1950, Kendall Papers, box 16, folder "Telegrams of Congratulation."

81. Leonard Engel, "Cortisone and Plenty of It," *Harper's,* September 1951, 56.

82. Joseph Monachino, "Strophanthus, Sarmentogenin, and Cortisone: The Botanical Aspects of the Story of the Newest 'Miracle Drug,'" *Journal of the New York Botanical Garden* 51, no. 602 (1950): 25–39.

83. Joseph Monachino, "Recent Developments in Strophanthus as a Precursor of Cortisone," *Journal of the New York Botanical Garden* 51, no. 610 (1950): 223–41.

84. Howard W. Blakeslee, "New Era in Medicine Launched," *Ogden (UT) Standard-Examiner,* February 4, 1951.

85. *Annual Report of the Federal Security Agency 1950* (Washington, DC: Government Printing Office, 1951).

86. "Health Aid Listed in Federal Report," *New York Times,* April 15, 1951.

87. G. Hetenyi Jr. and J. Karsh, "Cortisone Therapy: A Challenge to Academic Medicine in 1949–1952," *Perspectives in Biology and Medicine* 40, no. 3 (1997): 426–39.

88. D. S. Correll et al., "The Search for Plant Precursors of Cortisone," *Economic Botany* 9, no. 4 (October–December 1955): 307–75.

89. Leonard Engel, *Medicine Makers of Kalamazoo* (New York: McGraw-Hill, 1961).

90. Kreig, *Green Medicine,* 66.

10. The Hell Bomb

1. Joseph and Stewart Alsop, "Matter of Fact," *Washington Post,* January 18, 1950; Joseph and Stewart Alsop, "Matter of Fact," *Washington Post,* January 6, 1950; Joseph and Stewart Alsop, "Matter of Fact," *Washington Post,* January 4, 1950; Joseph and Stewart Alsop, "Matter of Fact," *Washington Post,* January 2, 1950.

2. James Reston, "U.S. Hydrogen Bomb Delay Urged Pending Bid to Soviet," *New York Times,* January 17, 1950.

3. "Fateful Issue Rises over Hydrogen Bomb," *New York Times,* January 9, 1950; William L. Laurence, "Much Hydrogen Bomb Data Known; Process Involves Fusion of Atoms," *New York Times,* January 18, 1950.

4. William L. Laurence, "Scientists Willing to Build New Bomb," *New York Times,* January 26, 1950.

5. William L. Laurence, "12 Physicists Ask U.S. Not to Be First to Use Super Bomb," *New York Times,* February 5, 1950.

6. William Leonard Laurence, interview by Scott Bruns, Columbia Center for Oral History, Columbia University, New York.

7. "People Are Talking about . . ." *Vogue,* September 1951, 186.

8. William L. Laurence, "It's a Triton Bomb, Mightiest Possible," *New York Times,* February 1, 1950.

9.. "Here It Is! . . . Secret of Hydrogen Bomb," *The Era* (Bradford, PA), February 11, 1950; " 'Triton'—not 'Hydrogen,' " *Sydney (Australia) Morning Herald,* February 2, 1950.

10. "H-Bomb Called 'Monstrosity' Able to 'Destroy Mankind,' " *Baltimore Sun,* February 13, 1950.

11. Hans Bethe et al., "The Facts about the Hydrogen Bomb," *Bulletin of the Atomic Scientists* 6, no. 4 (April 1950): 106–9.

12. "Four Scientists View H-Bomb as Suicide Weapon," *Chicago Daily Tribune,* February 27, 1950; "Hydrogen Hysteria," *Time,* March 6, 1950, 90.

13. William L. Laurence, "The Truth About the Hydrogen Bomb," *Saturday Evening Post,* June 24, 1950, 90.

14. Contract between National Broadcasting Company and William Laurence, May 24, 1950, and program information from NBC audience promotion for *The Quick and the Dead,* NBC files, box 289, folder 10.

15. Matthew C. Ehrlich, "Living with the Bomb," *Journalism History* 35, no. 1 (Spring 2009): 2–11.

16. NBC News and Special Events Department Presents "The Quick and the Dead" (Program 1 of 2), Friendly Papers, box 179, folder "The Quick and the Dead—Program transcripts," 2, 4.

17. June Bundy, "The Quick and the Dead," *Billboard,* July 15, 1950, 8.

18. "Radio Review: The Quick and the Dead," *Variety,* July 12, 1950, 30.

19. "NBC Will Rebroadcast Series for First Time," *Radio Daily,* August 4, 1950, 1.

20. Memorandum from Fred Friendly to Messrs. Brooks and Meyers, August 1, 1950, NBC files, box 290, folder 35.

21. Edward Tatnall Canby, "The New Recordings," *Saturday Review,* June 23, 1951, 47.

22. Warren Cheney, "What Is Happening in Civil Defense Films?" *Business Screen Magazine* 12, no. 5 (1951): 22.

23. *Pattern for Survival,* directed by George Carillon (Cornell Films, 1951), posted online by Periscope Film at https://www.youtube.com/channel/UCddem5RlB3bQe99wyY49g0g.

24. David Holbrook Culbert, Richard E. Wood, and Lawrence H. Suid, *Film and Propaganda in America: A Documentary History,* vol. 4 (New York: Greenwood, 1990).

25. "Individual Protection against Atom Bomb Depicted in a New 21-Minute Color Film," *New York Times*, December 21, 1950.

26. "Shorts Reviews: Pattern for Survival," *Boxoffice*, November 4, 1950.

27. Cecile Starr, "Ideas on Film," *Saturday Review*, December 1, 1952.

28. William L. Laurence, "U.S. Atomic Defense Guide Lists Steps for Saving Lives," *New York Times*, August 13, 1950; Laurence interview with Bruns.

29. Contract between Alfred A. Knopf Inc. and William L. Laurence, February 9, 1950, Knopf Records, box 88.7.

30. Correspondence from Alan C. Collins to Blanche W. Knopf, June 26, 1950, and BWK [Blanche W. Knopf], Memo for Files, June 23, 1950, both in Knopf Records, box 88.7.

31. Correspondence from Alfred A. Knopf to Harold Strauss, August 2, 1950; correspondence from Harold Strauss to William L. Laurence, August 16, 1950; correspondence from Harold Strauss to Alan C. Collins, September 6, 1950; correspondence from Harold Strauss to Alan C. Collins, August 28, 1950; and correspondence from Harold Strauss to Alan C. Collins, September 8, 1950, all in Knopf Records, box 88.7.

32. Fletcher Pratt, Reader Report on 'The Hell Bomb,'" and Donald Armstrong, Reader Report on 'The Hell Bomb,'" both in Book-of-the-Month Club Records, box 44, folder "1951 H."

33. Donna Hendleman, "H-Bomb Should Guarantee World Peace," *Michigan Daily*, November 30, 1950.

34. "Red Atomic Weapons Called 'Jalopy' Bombs," *Edwardsville (IL) Intelligencer*, January 23, 1951.

35. "A 'Defrighten' Campaign Prescribed," *The Bee* (Danville, VA), April 5, 1951; "Process Industries' Outlook," *Chemical and Engineering News* 29, no. 15 (1951): 1454–55; J. Frank Beatty, "Rear-Guard Attack," *Broadcasting, Telecasting*, April 2, 1951.

36. Postcard from John E. Kiernan to N.Y. Times Editorial Desk, January 25, 1951, Catledge Papers, series IIA, folder "Laurence William L., 1951."

37. Howard W. Blakeslee, "'Hell Bomb' Possible Science Writer Says, Warning of New Horrors," January 4, 1951, AP archives.

38. James R. Newman, "Books in Review," *New Republic*, January 22, 1951, 18.

39. "Briefly Noted," *New Yorker*, January 27, 1951, 79.

40. "What about the H-Bomb?" *Life*, January 8, 1951, 60.

41. G. P. Thomson, "The H-Bomb," *New Statesman and Nation*, May 26, 1951, 600.

42. William A. Higinbotham, "The H-Bomb and the Clouded Future," *New York Times*, January 7, 1951.

43. "Books," *Engineering and Science*, February 1951, 4.

44. Correspondence from Paul Palmer to Robert F. Bacher, January 10, 1951, and correspondence from Paul Palmer to Robert F. Bacher, January 11, 1951, both in Bacher Papers, box 37, folder 11.

45. Correspondence from Robert F. Bacher to Paul Palmer, January 18, 1951, Bacher Papers, box 37, folder 11.

46. Correspondence from William A. Koshland to Sewell Haggard, December 5, 1950, correspondence from William A. Koshland to William L. Laurence, February 26, 1951, and correspondence from Harold Strauss to William L. Laurence, January 26, 1951, all in Knopf Records, box 88.7; William L. Laurence, "Do We Want the H-Bomb?" *Catholic Digest*, December 1951, 102.

47. Lansing Lamont, interview with William L. Laurence, Lamont Papers, box 1, folder 2.

48. William L. Laurence, "Soviet Gain on U.S. in Bomb Race Seen," *New York Times*, February 4, 1950.

49. Laurence, "The Truth about the Hydrogen Bomb," 18.

50. William L. Laurence, "Fuchs Gave Soviet the Secret of Hydrogen Bomb in 1944," *New York Times,* August 9, 1953.

51. Bob Considine, "FBI Spy Tracks Down Harry Gold," *Salt Lake Tribune,* December 19, 1951; Ferenc Morton Szasz, *British Scientists and the Manhattan Project: The Los Alamos Years* (New York: St. Martin's Press, 1992), 94.

52. Waldemar Kaempffert, "Major Questions about the H-Bomb," *New York Times,* August 16, 1953.

53. Marquis Childs, "Atomic Program Is Making Fast Gains," *Washington Post,* June 9, 1953; Edward F. Ryan, "H-Bomb 'Delay' Seems Hard to Pin Down," *Washington Post and Times Herald,* May 2, 1954.

54. James R. Shepley and Clay Blair Jr., *The Hydrogen Bomb: The Men, the Menace, the Mechanism* (New York: D. McKay, 1954).

55. William J. Broad and Walter Sullivan, "Edward Teller, a Fierce Architect of the Hydrogen Bomb, Is Dead at 95," *New York Times,* September 11, 2003.

11. Atomic Dialogue

1. William L. Laurence, "Tests of Atomic Artillery Indicated; Greatest Nevada Blast Lights West," *New York Times,* February 7, 1951.

2. "Secrecy Shrouds Pace Trip to Pacific," *Lowell (MA) Sunday Sun,* April 8, 1951.

3. William L. Laurence, "Design Stage Seen on Hydrogen Bomb," *New York Times,* May 26, 1951; Austin Stevens, "U.S. Hints Progress on Hydrogen Bomb in Eniwetok Tests," *New York Times,* May 26, 1951.

4. Roy B. Snapp, "Admission of Limited, Selected Group of Uncleared Observers to One Shot Spring 1952 Test Series in Nevada," March 11, 1952, Energy Department archive, accession no. NV0030928.

5. "AEC v. the Reporters," *Time,* November 12, 1951, 42.

6. William L. Laurence, "Desert Blast Held 'Babybomb's' Debut," *New York Times,* October 24, 1951.

7. Gladwin Hill, "Atomic Blast a Mile Wide Set Off from Air in Nevada," *New York Times,* October 31, 1951; Gladwin Hill, "Atom Test Put Off at Final Moment," *New York Times,* October 27, 1951; Gladwin Hill, "Huge Blast Marks First Atom Games Involving Troops," *New York Times,* November 2, 1951; Gladwin Hill, "Atom Battle Bomb Dims Sun in Nevada," *New York Times,* October 29, 1951.

8. "Laurence Says AEC Press Policy Too Rigid," *Editor and Publisher,* November 24, 1951, 22.

9. Snapp, "Admission of Limited, Selected Group."

10. William L. Laurence, "Atom Blast Today to Be Strongest of All Except Those at Eniwetok," *New York Times,* April 22, 1952.

11. Klaus Landsberg, *Operation Big Shot,* UCLA Archive, videocassette VA2161 T.

12. William L. Laurence, "Atom Bomb Fired with Troops Near; Chutists Join Test," *New York Times,* April 23, 1952.

13. Marcel Chotkowski LaFollette, *Science on American Television: A History* (Chicago: University of Chicago Press, 2013), 217.

14. "Purpose of Nevada Atomic Test," *Times* (London), April 24, 1952; Hill Williams, "'My First Atomic Explosion,'" *Seattle Times Magazine,* August 6, 1972.

15. U.S. Department of Energy, *United States Nuclear Tests, July 1945 through September 1992* (Las Vegas: US Department of Energy Nevada Operations Office, 2000), http://www.nv.doe.gov/library/publications/historical/DOENV_209_REV15.pdf.

16. William L. Laurence, "Russian Progress in the Bomb Slow," *New York Times*, October 4, 1951.

17. Hanson W. Baldwin, "Soviet Atomic Power," *New York Times*, October 5, 1951.

18. Robert S. Norris and Hans M. Kristensen, "Global Nuclear Weapons Inventories, 1945–2010," *Bulletin of the Atomic Scientists* 66, no. 4 (July/August 2010): 77–83.

19. Memorandum from Hanson Baldwin to Mr. Catledge, October 4, 1951, Baldwin Papers, box 11, folder 556.

20. Baldwin, "Soviet Atomic Power," 13.

21. Memorandum from Turner Catledge to Mr. Baldwin, October 5, 1951, Baldwin Papers, box 11, folder 556.

22. Memorandum for Mr. Catledge, April 12, 1952, Catledge Papers, series IIA, folder "Laurence William L., 1951."

23. Chuck Hansen, *U.S. Nuclear Weapons: The Secret History* (Arlington, TX: Aerofax, 1988).

24. Stewart Alsop, " 'Hydrogen Era' Is Reds' Chance," *Washington Post*, November 16, 1952.

25. "Hydrogen Bomb Gains Power through Fusion," *New York Times*, November 17, 1952.

26. William L. Laurence, "Hydrogen Is Fused for Peace or War," *New York Times*, January 5, 1953.

27. William L. Laurence, "Vast Atom Strides Expected in Tests," *New York Times*, March 15, 1953.

28. William L. Laurence, "Atom Blast Today a Tactical Weapon," *New York Times*, March 17, 1953.

29. William L. Laurence, "Millions on TV See Explosion That Rocks Desert Like Quake," *New York Times*, March 18, 1953.

30. Correspondence from Alfred A. Knopf to Alan Collins, April 16, 1952, and correspondence from Alan C. Collins to Alfred A. Knopf, April 9, 1952, both in Knopf Records, box 108, folder 6.

31. William L. Laurence, "How Hellish Is the H-Bomb?," *Look*, April 21, 1953, 31.

32. Laurence, "How Hellish Is the H-Bomb?," 35.

33. William L. Laurence, "Atom Scientist Develops TV Tube Giving Both Color and Monochrome," *New York Times*, September 20, 1951.

34. "IA-Par Deal Key to SPG Future," *Variety*, September 26, 1955; Albert D. Hughes, "New Tube May Clear Up TV Color Rift," *Christian Science Monitor*, September 20, 1951; "New TV Color Tube Gives 'Excellent Results' in Tests," *Burlington (IA) Hawk Eye Gazette*, September 21, 1951.

35. Michael A. Hiltzik, *Big Science: Ernest Lawrence and the Invention That Launched the Military-Industrial Complex* (New York: Simon & Schuster, 2015).

36. William L. Laurence, "You May Live Forever," *Look*, March 24, 1953, 29.

37. E. B. White, "Notes and Comment," *New Yorker*, March 28, 1953, 23.

38. For example, "Forecast of Immortality Is Wondrous—and Frightening," *Logansport (IN) Press*, April 4, 1953; "Forecast of Immortality," *Sunday Messenger* (Athens, OH), April 19, 1953; "The Forecast of Immortality," *Evening Standard* (Uniontown, PA), April 7, 1953; Bruce Biossat, "Forecast of Immortality Wondrous but Frightening," *Austin (MN) Daily Herald*, April 4, 1953.

39. D. A. Delafield, "The Beginnings of a Dangerous New Science," *Advent Review and Sabbath Herald*, April 23, 1953.

40. Correspondence from Leland Hayward to Gordon Dean, May 11, 1953, Hayward Papers, box 111, folder 4.

41. Correspondence from Mrs. James Byrne to Lawrence White, June 25, 1953, Hayward Papers, box 111, folder 4.

42. Correspondence from Sherman Gross to Larry White, July 14, 1953, Hayward Papers, box 111, folder 4.

43. William L. Laurence, "Atomic Dialogue (Rough Preliminary Sketch)," Hayward Papers, box 112, folder 3.

44. William L. Laurence, "Atomic Dialogue," May 21, 1953, Hayward Papers, box 112, folder 3.

45. Jack Gould, "Television in Review," *New York Times,* June 16, 1953.

46. "Radio-Television: Inside Stuff-Television," *Variety,* June 24, 1953, 44.

47. John Frederick, "Ford Takes Over 2 Hours on TV Networks Tonight," *Long Beach (CA) Press-Telegram,* June 15, 1953; "2-Hour Show to Review Past 50 Years," *Charleston (WV) Daily Mail,* June 14, 1953; Terry Vernon, "TV Tele-Vues," *Long Beach (CA) Independent,* June 15, 1953; Ellis Walker, "Video Notes," *Hayward (CA) Daily Review,* June 9, 1953.

48. Correspondence from Alan C. Collins to Lawrence White, June 18, 1953, Hayward Papers, box 111, folder 4.

12. The U-Bomb

1. "Atom Blast Opens Test in Pacific; No Hint of Hydrogen Plans Given," *New York Times,* March 2, 1954.

2. "264 Exposed to Atom Radiation after Nuclear Blast in Pacific," *New York Times,* March 12, 1954.

3. Elie Abel, "Hydrogen Blast Astonished Scientists, Eisenhower Says," *New York Times,* March 25, 1954.

4. "New Hydrogen Explosion Is Set Off in Pacific Tests," *New York Times,* March 30, 1954.

5. William L. Laurence, "H-Bomb Can Wipe Out Any City, Strauss Reports after Tests; U.S. Restudies Plant Dispersal," *New York Times,* April 1, 1954, 1.

6 Robert A. Divine, *Blowing on the Wind: The Nuclear Test Ban Debate, 1954–1960* (New York: Oxford University Press), 13.

7. Robert H. Ferrell, *The Diary of James C. Hagerty: Eisenhower in Mid-Course, 1954–1955* (Bloomington: Indiana University Press, 1983).

8. "Color Film of First H-Bomb Test Is Previewed by Press in Capitol," *New York Times,* April 1, 1954, 1.

9. "Pearson 'Scoop' Cracks H-Bomb Film Embargo," *Editor and Publisher,* April 3, 1954, 11.

10. "Pearson Denies Embargo Break on H-Bomb Film," *Sandusky (OH) Register,* April 2, 1954, 2.

11. William L. Laurence, "Pacific Bomb Tests Hurled Man into the Hydrogen Age," *New York Times,* April 2, 1954.

12. William L. Laurence, "Tests Catapult Science from Atomic to Hydrogen Age in Giant Step," *Corpus Christi (TX) Times,* April 3, 1954.

13. "'Old Sloppy' Big Surprise," *Record-Eagle* (Traverse City, MI), April 12, 1954; "Cheap Method of Producing H-Bombs Reported," *Daily Redlands (CA) Facts,* April 1, 1954.

14. William H. Stringer, "U.S. Builds Simple H-Bomb Based upon Soviet Model," *Christian Science Monitor,* April 1, 1954, 1.

15. Richard Rhodes, *Dark Sun: The Making of the Hydrogen Bomb* (New York: Simon & Schuster, 1995).

16. " 'Ostrich-Like' U.S. Attitude on A-Secrets Rapped," *Ellensburg (WA) Daily Record*, March 17, 1954.

17. Memorandum from A. H. Belmont to L. V. Boardman, April 5, 1954, Laurence FBI file, 2.

18. Richard G. Hewlett and Jack M. Holl, *Atoms for War and Peace, 1953–1961* (Berkeley: University of California Press, 1989), 68.

19. "Columbia to Give 33 Honor Degrees," *New York Times*, May 20, 1954; telegram from Florence and Bill Laurence to Lewis L. Strauss, Strauss Papers, series 5, box 509, folder 7.

20. Correspondence from Florence Laurence to Alice and Lewis Strauss, Strauss Papers, series 5, box 509, folder 7.

21. Correspondence from Florence Laurence to Lewis Strauss, March 25, 1957; correspondence from Lewis L. Strauss to Mrs. William L. Laurence, March 20, 1957, Strauss Papers, series 5, box 509, folder 7.

22. Lewis L. Strauss telephone log, April 2–6, 1954, AEC archives, entry 10, box 5, folder "Telephone Log for LLS (March and April 1954)"; Lewis L. Strauss appointment calendar, April 9, 1954, AEC archives, entry 10, box 6, folder "Appointment Calendar LLS Jan.–June 1954."

23. Belmont memorandum to Boardman.

24. Minutes, Thirty-Ninth Meeting of the General Advisory Committee to the US Atomic Energy Commission, March 31, April 1 and 2, 1954, Washington, DC, Energy Department archives, document no. AECGAC39, 36.

25. William L. Laurence, "Fission Bomb Is Obsolete Except as Fusion Trigger," *New York Times*, April 3, 1954.

26. Belmont memorandum to Boardman.

27. Correspondence from K. D. Nichols to J. Edgar Hoover, April 6, 1954, FBI file 117-1686, obtained under Freedom of Information Act.

28. William L. Laurence, "Now Most Dreaded Weapon, Cobalt Bomb, Can Be Built," *New York Times*, April 7, 1954.

29. Memorandum from Shelby Thompson to Lewis L. Strauss, Henry D. Smith, Thomas E. Murray, Eugene M. Zuckert, Joseph Campbell, and K. D. Nichols, April 7, 1954, Strauss Papers, box 509, folder 7.

30. Lewis L. Strauss telephone log, April 2–6, 1954; Lewis L. Strauss appointment calendar, April 9, 1954.

31. Memorandum from FBI Director to Attorney General, April 13, 1954, William L. Laurence FBI file.

32. Kenneth D. Nichols diary, April 9, 1954, Nichols Papers, folder 79, 132.

33. Minutes, Fortieth Meeting of the General Advisory Committee to the U.S. Atomic Energy Commission, May 27, 28, and 29, 1954, Washington D.C., Energy Department archives, document no. AECGAC40, 4.

34. Memorandum from A. H. Belmont to L.V. Boardman, April 12, 1954, William L. Laurence FBI file, 4–5.

35. "Critic of Strauss to Quit Atom Post," *New York Times*, May 22, 1954.

36. US Department of Energy, Office of Health, Safety and Security, Office of Classification, *Restricted Data Declassification Decisions 1946 to the Present RDD-8* (Washington, DC: US Department of Energy, 2002), https://www.osti.gov/opennet/policy (accessed May 30, 2020).

37. Michiji Konuma, "Personal Contacts with Sir Joseph Rotbat," in *Joseph Rotblat: Visionary for Peace*, ed. Reiner Braun et al. (Weinheim, Germany: Wiley-VCH Verlag GmbH, 2007), 183–88.

38. Ralph E. Lapp, "Civil Defense Faces New Peril," *Bulletin of the Atomic Scientists*, November 1954, 349–51.

39. Ralph E. Lapp, "Radioactive Fall-Out," *Bulletin of the Atomic Scientists*, February 1955, 45; Ralph E. Lapp, " 'FALL-OUT' Another Dimension in Atomic Killing Power," *New Republic*, February 14, 1955, 8.

40. Edwin Diamond, "Reveal 1954 Test of Super U-Bomb," *Milwaukee Sentinel*, March 6, 1955.

41. "AEC Silent on Report U.S. Has Devised Cheap 'U-Bomb,' " *Baltimore Sun*, March 7, 1955.

42. "Strauss Says U.S. Leads in Atomic Race," *Washington Post*, April 4, 1955.

43. "U-Bomb," *Time*, March 28, 1955, 66; "The Great Leveler," *Newsweek*, March 21, 1955, 62.

44. "Transcript of Presidential Press Conference on Foreign and Domestic Affairs," *New York Times*, March 17, 1955.

45. Memoranda from William L. Laurence to Mr. Schwarz, March 23 and 24, 1955, Catledge Papers, series IIA, folder "Laurence William L., 1955."

46. Willard F. Libby, "Radioactive Fall-Out," *Bulletin of the Atomic Scientists* 11, no. 7 (September 1955): 256–60.

47. "Atom Tests Called Harmless," *New York Times*, June 4, 1955.

48. Robert A. Divine, *Blowing on the Wind: The Nuclear Test Ban Debate, 1954–1960* (New York: Oxford University Press, 1978).

49. Anthony Leviero, "Cheaper H-Bomb Is Now Possible," *New York Times*, June 12, 1955.

50. Ralph E. Lapp, "Radioactive Fall-Out III," *Bulletin of the Atomic Scientists* 11, no. 6 (1955): 206–30.

51. William L. Laurence, "The H-Bomb Danger," *New York Times*, July 10, 1955, 24.

52. William C. Davidon, "Scientists' Warning Defined," *New York Times*, July 31, 1955.

53. William L. Laurence, "Murray Favors Showing World an H-Bomb Blast," *New York Times*, November 18, 1955.

54. "H-Bomb Blast Urged to Spur Peace Effort," *Los Angeles Times*, November 18, 1955.

55. Memorandum from R. E. Garst to Turner Catledge, March 8, 1956, Garst Papers, box 2, folder 34, "Laurence, William L. 1956–1957."

56. Gene Marine, "Delayed U-Bomb and the N.Y. Times," *Nation*, January 28, 1956.

57. Charles A. Sprague, "It Seems to Me," *Statesman Journal* (Salem, OR), February 1, 1956.

58. Jesse Zel Lurie, "Bikini Mystery," *Nation*, February 25, 1956.

59. Correspondence from Irving Morrissett to Arthur Hays Sulzberger, January 28, 1956, and correspondence from Arthur Hays Sulzberger to Irving Morrissett, January 31, 1956, both in Sulzberger Papers, box 109, folder 4.

60. Memorandum from William L. Laurence to Mr. Catledge, February 4, 1956, Catledge Papers, series IIA, folder "Laurence William L., 1956."

61. Memorandum from Arthur Hays Sulzberger to Mr. Catledge, February 11, 1956, Sulzberger Papers, box 109, folder 2.

62. Memorandum from R. E. Garst to Hanson Baldwin, March 2, 1956, Garst Papers, box 2, folder 34, "Laurence, William L. 1956–1957."

63. Memorandum from Hanson Baldwin to Mr. Garst, March 5, 1956, Catledge Papers, series IIA, folder "Laurence William L., 1955."

64. Garst to Catledge, March 8, 1956.

65. Memorandum from Leviero to Mr. Reston, February 21, 1956, Catledge Papers, series IIA, folder "Laurence William L., 1956."

66. Kosta Tsipis, *Arsenal: Understanding Weapons in a Nuclear Age* (New York: Simon & Schuster, 1983); Jeremy I. Pfeffer and Shlomo Nir, *Modern Physics: An Introductory Text* (London: Imperial College Press, 2000); Thomas B. Cochran, William M. Arkin, and Milton M. Hoenig, *Nuclear Weapons Databook*, vol. 1 (Cambridge, MA: Ballinger, 1984).

67. Ann Jacob Stocker, "A Study of the Life and Work of William L. Laurence as Related to His Contributions in Science Journalism" (master's thesis, Syracuse University, 1966).

68. Gladwin Hill, "Unusual Device Shows a 'Flaming Curtain' Plus Normal Fireball," *New York Times*, April 2, 1952; Gladwin Hill, "Huge Blast Marks First Atom Games Involving Troops," *New York Times*, November 2, 1951; Gladwin Hill, "Atom Battle Bomb Dims Sun in Nevada," *New York Times*, October 29, 1951; Gladwin Hill, "3d Atom Test Lights Nevada Dawn; Peaks Stand Out in Weird Glare," *New York Times*, February 2, 1951.

69. Beverly Deepe Keever, *News Zero: The New York Times and the Bomb* (Monroe, ME: Common Courage Press, 2004).

70. United States Naval Institute, *Reminiscences of Hanson Weightman Baldwin, U.S. Navy (Retired)* (Annapolis, MD: US Naval Institute, 1976), 535.

71. USNI, *Reminiscences*, 536.

72. Carroll Quigley, *Tragedy and Hope: A History of the World in Our Time* (New York: Macmillan, 1966), 967–68.

73. Defense Department and US Atomic Energy Commission, Joint Office of Test Information, Pooled Group of US Newsmen Invited to Cover Pacific Test (Release no. 1), Energy Department Archives, accession no. NV030511.

74. Defense Department and US Atomic Energy Commission, Joint Office of Test Information.

75. Correspondence from William L. Laurence to Lewis L. Strauss, July 18, 1956, and correspondence from Lewis L. Strauss to William L. Laurence, June 11, 1956, both in Strauss Papers, series 5, box 509, folder 7.

76. NARA crew lists, roll 11.

77. Darrell Garwood, "Newsmen to Cover H-Bomb Blast from 50 Miles Out," *Christian Science Monitor*, May 4, 1956.

78. William L. Laurence, "Atom Test Series Starts in Pacific," *New York Times*, May 5, 1956.

79. Joseph L. Myler, "Eniwetok Tests Seen Producing New H-Weapons," *Coshocton (OH) Tribune*, May 6, 1956; Joseph L. Myler, "Warhead Chief Goal of Tests," *Bakersfield Californian*, May 12, 1956.

80. US Department of Energy, *United States Nuclear Tests, July 1945 through September 1992* (Las Vegas: US Department of Energy Nevada Operations Office, 2015), https://www.nnss.gov/docs/docs_LibraryPublications/DOE_NV-209_Rev16.pdf (accessed May 29, 2020).

81. William L. Laurence, "H-Bomb Test First by Plane for U.S.," *New York Times*, May 6, 1956; US Department of Energy, *United States Nuclear Tests, July 1945 through September 1992*.

82. William L. Laurence, "Weather Delays Test for H-Bomb," *New York Times*, May 7, 1956; "Winds Shift Favorably," *New York Times*, May 10, 1956; "U.S. Again Postpones Test of Hydrogen Bomb Drop," *New York Times*, May 8, 1956.

83. Marvin Miles, "H-Blast May Be Twice as Strong as Expected," *Los Angeles Times*, May 7, 1956; Marvin Miles, "H-Bomb Blast Delayed Once Again by Weather," *Los Angeles Times*, May 13, 1956; Marvin Miles, "Unexpected Winds Again Delay Big H-Bomb Shot," *Los Angeles Times*, May 18, 1956; Marvin Miles, "Weather Delays H-Bomb Blast for Third Time," *Los Angeles Times*, May 9, 1956.

84. Joseph L. Myler, "U.S. to Explode New H-Bomb High in the Air Next Month," *Evening Chronicle* (Marshall, VA), April 26, 1956; Joseph L. Myler, " 'Public' H-Bomb Blast to Start Test Series," *Honolulu (HI) Advertiser*, April 27, 1956.

85. "Laurence Goes 15,000 Miles for Brief Look at H-Bomb," *Times Talk*, June 1956, Sulzberger Papers, box 11, folder 13, 4.

86. Correspondence from Charter Heslep to Margaret Heslep, May 7, 1956, Heslep Papers, box 1, folder 6, 1.

87. Correspondence from Charter Heslep to Margaret Heslep, May 10, 1956, Heslep Papers, box 1, folder 6.

88. Ancient and Honourable Society of H-Bomb Observers, Oregon State University Libraries Special Collections and Archives Research Center, http://scarc.library.oregonstate.edu/omeka/items/show/1056 (accessed May 30, 2020).

89. Correspondence from Charter Heslep to Margaret Heslep, May 10, 1956, Heslep Papers, box 1, folder 6.

90. Correspondence from Charter Heslep to Margaret Heslep, May 16, 1956, Heslep Papers, box 1, folder 6.

91. Correspondence from Charter Heslep to Margaret Heslep, May 18, 1956, Heslep Papers, box 1, folder 6; "H-Bomb Test Area Combed for Airman," *New York Times*, May 19, 1956; Marvin Miles, "Pilot Saved, One Missing as B-57 on H-Blast Mission Falls in Sea," *Los Angeles Times*, May 19, 1956.

92. William L. Laurence, "Small U.S. H-Bomb Believed Tested," *New York Times*, May 15, 1956.

93. "Operation Redwing," Nuclear Weapon Archive, last modified October 22, 1997, http://nuclearweaponarchive.org/Usa/Tests/Redwing.html (accessed May 30, 2020).

94. Cochran et al., *Nuclear Weapons Databook*.

95. William L. Laurence, "Airborne H-Bomb Exploded by U.S. over Pacific Isle," *New York Times*, May 21, 1956.

96. Correspondence from Charter Heslep to Margaret Heslep, May 21, 1956, Heslep Papers, box 1, folder 6, 1–2. Either Heslep is incorrect in stating that Laurence had declared that the test was 25 megatons or Laurence somewhat reduced his estimate before he submitted his story. In his published story, the reporter estimated the yield as between 15 and 20 megatons—still a gross overestimate compared to the actual value of 3.8 megatons.

97. Memorandum from R. E. Garst to Orvil Dryfoos, May 25, 1956, Garst Papers, box 2, folder 34, "Laurence, William L. 1956–1957."

98. U.S. Department of Energy, *United States Nuclear Tests, July 1945 through September 1992*.

99. Laurence, "Airborne H-Bomb Exploded," "Operation Redwing," Nuclear Weapon Archive, last modified October 22, 1997, http://nuclearweaponarchive.org/Usa/Tests/Red wing.html.

100. Laurence, "Airborne H-Bomb Exploded," 16.

101. Marvin Miles, "Times Man Eyewitness Story of H-Bomb Drop," *Los Angeles Times*, May 21, 1956.

102. Joseph L. Myler, "H-Bomb Blast Is Awesome Answer to Russian Thermonuclear Boast," *Brownwood (TX) Bulletin*, May 21, 1956.

103. Bob Considine, "H-Bomb Blast! World Feels it," *Daily Defender,* May 21, 1956.

104. Philip W. Porter, "Porter on Big Bombs," *Cleveland Plain Dealer,* May 26, 1956.

105. "Bill Laurence," *Newsletter of the National Association of Science Writers,* December 1956.

106. "Scientists Studying H-Bomb Blast Data," *New York Times,* May 22, 1956.

107. Keever, *News Zero.*

108. Correspondence from Shelby Thompson to Joseph L. Myler, June 8, 1956, AEC archives, box 183, folder "Military research & application."

109. AEC-MLC Minutes, 5–22–56, AEC archives, box 183, folder "Military research & application."

110. "Operation Redwing."

111. Laurence FBI file.

112. Garst to Dryfoos, May 25, 1956.

113. Correspondence from R. E. Lapp to Arthur Hays Sulzberger, May 28, 1956, Catledge Papers, series IIA, folder "Laurence William L., 1956."

114. Correspondence from Ralph E. Lapp to Evan Thomas, May 28, 1956, Harper Records, box 16, folder 30.

115. Memorandum from R. E. Garst, to Mr. Schwarz and Mr. Merz, May 29, 1956, Garst Papers, box 2, folder 34, "Laurence, William L. 1956–1957."

116. William L. Laurence, "H-Bomb Improved by Fall-Out Curb," *New York Times,* July 29, 1956.

117. Correspondence from C. A. Rolander Jr. to J. Edgar Hoover, August 23, 1956, FBI file 117–2071, obtained through Freedom of Information Act.

118. Correspondence from J. A. Waters to J. Edgar Hoover, November 15, 1956, FBI file 117–2071, obtained through Freedom of Information Act, 2.

119. Correspondence from William F. Tompkins to Director, Federal Bureau of Investigation, November 27, 1956, FBI file 117–2071, obtained through Freedom of Information Act.

13. King Laurence

1. "Collier's Tells You Why [Advertisement]," *Boston Traveler,* May 24, 1956.

2. "Sunday [Advertisement]," *New York Journal-American,* July 27 and 28, 1956, and "Today's Exclusive Features," *New York Journal-American,* July 29, 1956.

3. "Today's Exclusive Features," 8, 23.

4. Memorandum from Arthur Hays Sulzberger to Mr. Dryfoos, July 30, 1956, and untitled memorandum, July 27, 1956, both in Sulzberger Papers, box 41, folder 24.

5. Memorandum from Lester Markel to Mr. Catledge, August 13, 1956, Catledge Papers, series IIA, folder "Laurence William L., 1956."

6. William L. Laurence, "Why There Can Not Be Another War," *Reader's Digest,* November 1956.

7. "Julie Andrews Says: "Verily, a Professor 'Iggins among Magazines! [Advertisement]," *Yellow Jacket* (Brownwood, TX), November 6, 1956.

8. Memorandum from Orvil E. Dryfoos to Mr. Sulzberger, December 11, 1956, Sulzberger Papers, box 41, folder 24; Employment Record for Laurence, William L., Catledge Papers, series IIC, box 5, folder "Laurence William L., 1952–1968."

9. J. Desmond, memorandum from Sunday Department, January 9, 1957, Markel Papers, box 20, folder 16, "Science news 1956–1962."

10. J. Desmond, memorandum from Sunday Department, February 8, 1957, Markel Papers, box 20, folder 16, "Science news 1956–1962."

11. R. G. Whalen, memorandum from Sunday Department, March 15, 1957, Markel Papers, box 20, folder 16, "Science news 1956–1962."

12. J. Desmond, Memorandum from Sunday Department, May 8, 1957, Markel Papers, box 20, folder 16, "Science news 1956–1962."

13. Memorandum from William L. Laurence, July 1957, Foreign Desk Records, box 60, folder 14, "Laurence, William L. 1957."

14. Memorandum from Turner Catledge to Orvil Dryfoos, July 4, 1957, Foreign Desk Records, box 60, folder 14, "Laurence, William L. 1957."

15. Memorandum from R. E. Garst to Mr. Catledge, July 10, 1957, Foreign Desk Records, box 60, folder 14, "Laurence, William L. 1957."

16. Memorandum from R. E. Garst to Mr. Glassberg, July 25, 1957, Garst Papers, box 2, folder 34, "Laurence, William L. 1956–1957."

17. Telegrams from Laurence to Nathaniel M. Gerstenzang; Bill Laurence to Reston, August 17, 1957; and Gerstenzang to Laurence, August 19, 1957, all in Catledge Papers, series IIA, folder "Laurence William L., 1957."

18. William L. Laurence, "Atom Physicists Meet in Israel," *New York Times*, September 10, 1957; William L. Laurence, "Gamma Rays Aid Studies of Atom," *New York Times*, September 11, 1957; William L. Laurence, "Physicists Hunt 2 New Particles," *New York Times*, September 13, 1957.

19. Telegram from Nathaniel M. Gerstenzang to London bureau, August 9, 1957; Catledge Papers, series IIA, folder "Laurence William L., 1957."

20. Memorandum from R. E. Garst to Turner Catledge, November 29, 1957, Garst Papers, box 2, folder 34, "Laurence, William L. 1956–1957."

21. William L. Laurence, *Science in Israel*, Herzl Institute Pamphlet, 4th ed. (New York: Herzl Institute, 1958).

22. Laurence, *Science in Israel*, 15.

23. CBS Television Network, "'The Twentieth Century' Presents 'Hiroshima,' March 9; Story of Secret Training and Flight of B-29 Crew That Dropped Atom Bomb," February 27, 1958, Benjamin Papers, box 1, folder 20, 2.

24. "Hiroshima Is Show Subject," *Amarillo (TX) Sunday News-Globe*, March 9, 1958.

25. "Story of Hiroshima," *Lima (OH) News*, July 18, 1959; CBS Television Network, "'The Twentieth Century' Presents 'Hiroshima.'"

26. J. Desmond, memorandum from Sunday Department, September 4, 1957, Markel Papers, box 20, folder 16, "Science news 1956–1962."

27. J. Desmond, memorandum from Sunday Department, December 28, 1957, Markel Papers, box 20, folder 16, "Science news 1956–1962."

28. J. Desmond, memorandum from Sunday Department, April 25, 1958, Markel Papers, box 20, folder 16, "Science news 1956–1962."

29. J. Desmond, memorandum from Sunday Department, May 27, 1958, Markel Papers, box 20, folder 16, "Science news 1956–1962."

30. J. Desmond, memorandum from Sunday Department, January 15, 1959, Markel Papers, box 20, folder 16, "Science news 1956–1962."

31. J. Desmond, memorandum from Sunday Department, December 8, 1959, Markel Papers, box 20, folder 16, "Science news 1956–1962."

32. J. Desmond, memorandum from Sunday Department, May 22, 1957, Markel Papers, box 14, folder 28, "International Geophysical Year 1957–1958"; William L. Laurence, "2 Rocket Experts Argue 'Moon' Plan," *New York Times*, October 14, 1952.

33. "Noted Author Discusses Red Satellite," *Daily Utah Chronicle* (University of Utah), October 11, 1957.

34. "Importance of Satellite Deflated by Discussion Club Science Speaker," *Newport (RI) Daily News,* December 13, 1957, 22.

35. William L. Laurence, "To Mars and Venus," *New York Times,* June 12, 1960, E9.

36. William L. Laurence, "Science—Our Heritage and Our Future," *Journal of the Franklin Institute* 273, no. 5 (1962): 361.

37. Joseph E. Seagram and Sons, Inc., "Life in Other Worlds" (1961), 41, Seagram records, box 876, folder "Industrial Relations, Miscellaneous, B—1948–1962."

38. Seagram and Sons, "Life in Other Worlds," 44.

39. William L. Laurence, "Science," *New York Times,* March 5, 1961.

40. Correspondence from Arthur Luce Klein to William L. Laurence, Spoken Arts Collection, box 11, folder 159.

41. Royalty statements, Spoken Arts Collection, box 35, folder 553.

42. William L. Laurence, *The Conquest of Space,* Spoken Arts SA 775 (1960), 33-1/3 rpm.

43. "Reviews of New Albums," *Billboard,* November 13, 1961, 44.

44. "Spoken Arts [Advertisement]," *Cornell Daily Sun,* December 11, 1964, 6.

45. Michael Korda, *Another Life: A Memoir of Other People* (New York: Random House, 1999), 77–78.

46. Robert H. Ferrell, *Harry S. Truman: A Life* (Columbia: University of Missouri Press, 1994), 349; Harry S. Truman and Ralph E. Weber, *Talking with Harry: Candid Conversations with President Harry S. Truman* (Wilmington, DE: SR Books, 2001).

47. John Barkham, "'Men and Atoms' Absorbing Story of Birth of A-Bomb," *Syracuse (NY) Herald-American,* October 11, 1959, 12; Christie, Manson & Woods International Inc., *The Personal Property of Marilyn Monroe: Auction Wednesday 27 and Thursday 28 October 1999* (New York: Christie's, 1999); Bob Considine, "A Silver Lady for Rube," *Lowell (MA) Sun,* November 12, 1959, 7; Mark Wolverton, "Books," *New Yorker,* October 17, 1959, 226–27.

48. "Guardian Savings [Advertisement]," *Press-Courier* (Oxnard, CA), September 2, 1961; "Strauss Favors Resuming Atomic Testing Program," *Albuquerque Journal,* August 28, 1961.

49. William L. Laurence, "The Neutron Bomb," *Saturday Evening Post,* May 5, 1962, 52.

50. "Will We Develop the Neutron Bomb? [Advertisement]," *Oakland (CA) Tribune,* May 1, 1962.

51. "What'll They Discover Next? [Advertisement]," *Sunday Gazette-Mail* (Charleston, WV), August 13, 1961.

52. Laurence FBI file.

53. Joseph A. Esposito, *Dinner in Camelot: The Night America's Greatest Scientists, Writers, and Scholars Partied at the Kennedy White House* (Lebanon, NH: ForeEdge, 2018).

54. Ann Jacob Stocker, "A Study of the Life and Work of William L. Laurence as Related to His Contributions in Science Journalism" (master's thesis, Syracuse University, 1966), 27.

14. Peace through Understanding

1. "Television Phone Used from Fair to California," *New York Times,* April 21, 1964.

2. New York State Power Authority, "Report on the Relationship of Nuclear Power to the St. Lawrence Power Project," St. Lawrence archives, box 66, 8.

3. Memorandum from J. Anthony Panuch to Mr. Moses, October 20, 1960, World's Fair records, box 456, folder "PR1.6 Laurence & Shuster Committee Luncheon—11/27/60, Luncheons & Banquets, Public Relations."

4. "Interim Report of the Executive Committee of the Fair to the President and Congress of the United States Suggesting for Consideration a Plan for a Permanent Franklin National Center of Science and Education to House the United States Exhibit at the Fair," United States Pavilion at the New York World's Fair, Summary of Important Events, http://www.worldsfairphotos.com/nywf64/documents/us-pavilion-important-events.pdf (accessed January 7, 2022), exhibit no. 2.

5. "U.S. Science Site Urged at '64 Fair," *New York Times,* December 5, 1960.

6. "Letter to Dwight D. Eisenhower, December 15, 1960," United States Pavilion at the New York World's Fair, Summary of Important Events, exhibit no. 3.

7. Correspondence from William L. Laurence to Jerome B. Wiesner, August 17, 1961, World's Fair records, box 255, folder "P0.0 Permanent Science Museum, Federal, Participation"; correspondence from Detlev W. Bronk to Jerome B. Wiesner, August 21, 1961, Bill Laurence to Detlev W. Bronk, August 16, 1961, and David Z. Beckler to Detlev W. Bronk, August 24, 1961, all in Bronk Papers, box 19, folder 15.

8. "Builder Appointed to Represent Federal Interests at 1964 Fair," *New York Times,* August 8, 1962.

9. Myron Cowen letter, February 8, 1961, World's Fair records, box 284, folder "P0.3 Task Force—Far East—Report."

10. Visit of New York World's Fair Committee to Phnom Penh, March 24, 1961, World's Fair records, box 284, folder "P0.3 Task Force—Far East—Report."

11. "Group to Aid Fair Here," *New York Times,* January 20, 1961.

12. T. F. Farrell, Atomic Plant and Desalting Plant, March 29, 1961, World's Fair records, box 284, folder "G3.35—Atomic Energy Plant, Production, Construction."

13. Abel Green, "Re-Define World's Fair 'Fun,'" *Variety,* November 22, 1961, 18.

14. Memorandum from Bill Berns to Robert Moses, November 13, 1961, World's Fair records, box 36, folder "A0.60 Federal Comm. Visit—11/16/61, Federal Legislation, Administration."

15. Correspondence from Robert Moses to William L. Laurence, September 24, 1962, World's Fair records, box 300, folder "P1.213—Space travel, air, Participation."

16. Memorandum from Robert Moses to Bill Berns, August 27, 1963, World's Fair records, box 396, folder "PR0.00 New York Times A-N, Newspapers, Public Relations (1959–1964)."

17. WABC Press Conference—10-16-63, World's Fair records, Audio tape control no. 01038.

18. "A Hall of Science," *New York Times,* April 24, 1963.

19. Robert Moses, Purpose of Hall of Science, April 24, 1963, World's Fair records, box 333, folder "P2.23—Hall of Science—1963".

20. "Calendar no. 190," *Journal of Proceedings of the Board of Estimate of the City of New York* 3 (1963): 2979–83.

21. "Officials Praise Science Hall Foe," *New York Times,* April 26, 1963.

22. Memorandum from John B. Oakes to Mr. Laurence, April 30, 1963, Oakes Papers, box 11, folder "William L. Laurence."

23. Memorandum from William L. Laurence to Mr. Oakes, May 3, 1963, Oakes Papers, box 11, folder "William L. Laurence."

24. Correspondence from Charles F. Preusse to John B. Oakes, October 11, 1963, World's Fair records, box 20, folder "Laurence, William L."

25. Letter to William L. Laurence, April 3, 1963, World's Fair records, box 20, folder "Laurence, William L."

26. Post Fair Expansion, Hall of Science, World's Fair records, box 31, folder 608.

27. Memorandum from William L. Laurence, October 15, 1963, Oakes Papers, box 11, folder "William L. Laurence."

28. Memorandum from A. O. Sulzberger to Mr. Cox, Mr. Markel, and Mr. Oakes, October 16, 1963, Oakes Papers, box 11, folder "William L. Laurence."

29. Gay Talese, *The Kingdom and the Power* (New York: World Publishing Co., 1969).

30. Talese, *The Kingdom and the Power*, 120.

31. Retirement Data Sheet, computed as of June 30, 1963, and memorandum from John B. Oakes for Mr. Lester Markel, October 7, 1963, both in Oakes Papers, box 11, folder "William L. Laurence"; memorandum from Sulzberger to Cox et al.

32. Guest list for Mr. William L. Laurence's Party—December 16, 1963, and correspondence from Bill Laurence to John [Oakes], December 20, 1963, both in Oakes Papers, box 11, folder "William L. Laurence."

33. "Science of Reporting," *Time*, December 27, 1963, 32.

34. Annotated copy of "Science of Reporting," *Time*, December 27, 1963, World's Fair records, box 20, folder "Laurence, William L."

35. William L. Laurence, "Hormone Found in Placenta Called Aid to Fetus," *New York Times*, January 1, 1964.

36. "W. L. Laurence of Times Retires," *New York Times*, January 1, 1964.

37. "Deegan Concern Elects," *New York Times*, November 29, 1968; "Mrs. William Berlinghoff, with the Times 40 Years," *New York Times*, January 22, 1968; "Laurence Appointed Aide to National Foundation," *New York Times*, April 1, 1964; "Laurence to Be Honored," *New York Times*, February 9, 1964; "Laurence Honored by Science Groups," *New York Times*, January 21, 1964.

38. Albert Rosenfeld, "Honor to the Dean," *Newsletter of the National Association of Science Writers*, March 1964.

39. Correspondence from Robert Moses to William L. Laurence, March 9, 1964, Moses Papers, box 49, folder "L."

40. William L. Laurence, "Alamogordo, Mon Amour," *Esquire*, May 1965, 118.

41. William L. Laurence, "Day of Trinity," *New York Times*, July 18, 1965; William L. Laurence, "The Start of It All," *New York Times*, April 2, 1967.

42. William L. Laurence, "Would You Make the Bomb Again?" *New York Times Magazine*, August 1, 1965, 8; William L. Laurence, "The Scientists: Their Views 20 Years Later," in *Hiroshima Plus 20* (New York: Delacorte Press, 1965), 114–25.

43. Memorandum from William Laurence to Robert Moses, May 14, 1964, World's Fair records, box 251, folder "M3.0 Hall of Science 1963–1963 (A-H) Permanent Buildings, Maintenance."

44. Post Fair Expansion, Hall of Science.

45. Robert Alden, "Science Museum May Be Expanded," *New York Times*, January 26, 1965.

46. Glenn T. Seaborg, *Journal of Glenn T. Seaborg, Chairman, U.S. Atomic Energy Commission, 1961–1971*, vol. 16 (Berkeley: Lawrence Berkeley Laboratory, University of California, 1989).

47. Bill Laurence, Post-Fair Exhibits for the Hall of Science, September 30, 1965, World's Fair records, box 333, folder "P2.23—Hall of Science—1965."

48. Sandra Blakeslee, "First Atomarium Planned Here: A Hot Reactor Open to Public," *New York Times*, January 7, 1969.

49. Richard J. H. Johnston, "Moses Is Elected President of Fair," *New York Times*, May 25, 1960.

50. Correspondence from Robert A. Harper to Robert Moses, May 11, 1966; memorandum from Robert Moses to Charles Preusse, May 9, 1966; memorandum from Robert Moses to Charles Preusse, April 1, 1966; correspondence from Robert Moses to Robert A. Harper, May 19, 1966, all in World's Fair records, box 333, folder "P2.23—Hall of Science—1966."

51. Walter Sullivan, "Hopeful Future Museum," *New York Times,* September 22, 1966; "Existing Structure," *New Yorker,* October 1, 1966, 41; Clayton Knowles, "New Science Hall Planned by City," *New York Times,* June 1, 1967; Corey Kilgannon, "The Rocket Park Reopens, and Cronkite Reminisces," *New York Times,* October 1, 2004; William Leonard Laurence, interview by Scott Bruns, Columbia Center for Oral History, Columbia University, New York, 539.

52. Beverly Ann Deepe Keever, *News Zero: The New York Times and the Bomb* (Monroe, ME: Commons Courage Press, 2004).

53. Leslie R. Groves, *Now It Can Be Told* (New York: Harper & Row, 1962), 326.

54. Correspondence from William L. Laurence to Arthur Hays Sulzberger, February 15, 1967, Sulzberger Papers, box 41, folder 24, 2.

55. Correspondence from Arthur Ochs Sulzberger to William L. Laurence, February 21, 1967, Catledge Papers, series IIC, box 5, folder "Laurence William L., 1952–1968."

56. Correspondence from Edward S. Greenbaum to Arthur O. Sulzberger, January 24, 1968, Sulzberger Papers, box 41, folder 24.

57. Turner Catledge, Memorandum for the File, January 12, 1968, Sulzberger Papers, box 41, folder 24.

58. Correspondence from Arthur Ochs Sulzberger to Edward S. Greenbaum, January 26, 1968, Sulzberger Papers, box 41, folder 24.

59. Correspondence from Edward S. Greenbaum to Arthur O. Sulzberger, February 5, 1968, Sulzberger Papers, box 41, folder 24.

60. Correspondence from Arthur Ochs Sulzberger to Edward S. Greenbaum, February 14, 1968, Sulzberger Papers, box 41, folder 24.

61. Jack O'Brian, "Beatles Planning to Open 'Discotek' in Manhattan," *Tonawanda (NY) News,* January 12, 1968.

62. R. J. Buswell, *Mallorca and Tourism* (Clevedon, UK: Channel View Publications, 2011).

63. H. P. Koenig, "Beautiful Mallorca," *Chicago Tribune,* January 2, 1977.

64. Curt L. Heymann, "News from Majorca," *Overseas Press Club Bulletin,* August 10, 1968, 7.

65. Correspondence from Florence Laurence to Robert and Mary Moses, November 6, 1968, Moses Papers, box 54, folder "L."

66. Correspondence from William L. Laurence to Robert Moses, February 24, 1969, Moses Papers, box 56, folder "L."

67. Correspondence from Florence Laurence to Mary and Robert Moses, August 18, 1969, Moses Papers, box 56, folder "L."

68. Correspondence from William L. Laurence to Robert Moses, June 29, 1970, Moses Papers, box 59, folder "L."

69. William L. Laurence, "Now We Are All Sons-of-Bitches," *Science News,* July 11, 1970, 39.

70. William L. Laurence, letter to the editor of the *New York Times,* December 2, 1971, Sulzberger Papers, box 11, folder 13.

71. Correspondence from Kalman Seigel to William L. Laurence, January 21, 1972, Sulzberger Papers, box 11, folder 13.

72. William Berkowitz, ed., *Let Us Reason Together* (New York: Crown, 1970), 16.

73. William L. Laurence, "Induced Biotin Deficiency as a Possible Explanation of Observed Spontaneous Recessions in Malignancy," *Science* 94, no. 2430 (July 25, 1941): 88–89.

74. "Biotin in Tumor Tissue," *Journal of the American Medical Association* 117, no. 8 (August 23, 1941): 622.

75. William L. Laurence, "Evidence in Support of Hypothesis Explaining Observed Spontaneous Recessions in Human Malignancy," Bronk Papers, box 10, folder "Laurence, William L. 1956–75."

76. Correspondence from William L. Laurence to Detlev W. Bronk, February 17, 1972, Bronk Papers, box 10, folder "Laurence, William L. 1956–75."

77. Correspondence from William Trager to Detlev W. Bronk, April 3, 1972, Bronk Papers, box 10, folder "Laurence, William L. 1956–75."

78. Correspondence from Detlev W. Bronk to William L. Laurence, May 3, 1972, Bronk Papers, box 10, folder "Laurence, William L. 1956–75."

79. Correspondence from William L. Laurence to Detlev W. Bronk, May 16, 1972, Bronk Papers, box 10, folder "Laurence, William L. 1956–75."

80. William L. Laurence, "La Causa de las Recesiones Espontaneas en los Tumores Malignos," *Archivos de la Facultad de Medicina de Madrid* 27, no. 3 (1975): 155–58.

81. Correspondence from Detlev W. Bronk to Mr. and Mrs. William Laurence, July 10, 1975, and Florence D. Laurence to Robert and Mary Moses, May 25, 1975, both in Bronk Papers, box 10, folder "Laurence, William L. 1956–75."

82. Correspondence from Florence D. Laurence to Robert and Mary Moses, January 24, 1973, Moses Papers, box 65, folder "K-L 1973."

83. Correspondence from Robert Moses to Mr. and Mrs. William L. Laurence, May 18, 1973, Moses Papers, box 65, folder "K-L 1973." Moses's article was Robert Moses, "Read Now, Fly Later," *Travel and Leisure,* Autumn 1973, FE2.

84. Correspondence from Florence D. Laurence to Robert and Mary Moses, January 30, 1974, Bronk Papers, box 10, folder "Laurence, William L. 1956–75."

85. Correspondence from Florence D. Laurence to Robert and Mary Moses, July 8, 1974, Bronk Papers, box 10, folder "Laurence, William L. 1956–75."

86. Correspondence from Florence D. Laurence to Robert and Mary Moses, Bronk Papers, box 10, folder "Laurence, William L. 1956–75."

87. Patricia Lee Brookhart, "Report on the Death of an American Citizen" for William Leonard Laurence, April 19, 1977, US State Department, obtained by author from the US State Department.

88. Correspondence from A. M. Rosenthal to Florence Laurence, March 25, 1977, Rosenthal Papers, box 25, folder 24, "Laurence, William L. 1977."

89. Correspondence from Florence Laurence to Robert and Mary Moses, September 12, 1979, Moses Papers, box 68, folder "L."

90. John Crewdson, "'Perky Cheerleaders,'" *Nieman Reports* 47, no. 4 (1993): 12.

91. William L. Laurence, "Journalistic Aspects of Science Writing," in *Writing in Industry*, vol. 1, ed. Siegfried Mandel (Brooklyn, NY: Polytechnic Press, 1959), 109.

92. William L. Laurence, "Eyewitness to a Later Day Genesis," Schuster Papers, box 274, folder "William L. Laurence."

93. William L. Laurence, "Science—Our Heritage and Our Future," *Journal of the Franklin Institute* 273, no. 5 (1962): 360.

94. Natalie Angier, "The Lab as a Battlefield," *New York Times,* March 24, 1991.

95. William J. Broad, "The Truth Behind the News," *New York Times*, August 10, 2021.

96. "William Laurence of the Times Dies," *New York Times,* March 19, 1977.

97. "William L. Laurence," *New York Times,* March 24, 1977.

98. "William L. Laurence, 89, Honored Science Writer," *Washington Post,* March 20, 1977.

99. Earl Ubell, "William Laurence: 1889–1978 [*sic*]," *Newsletter of the National Association of Science Writers* 26, no. 2 (1978): 15.

100. Brookhart, "Report on the Death of an American Citizen."

101. Memorandum from James M. Markham, March 21, 1977, Rosenthal Papers, box 25, folder 24, "Laurence, William L. 1977."

Note on Sources

Archival sources have been key in tracking Laurence's work and interactions over the decades. The National Archives holdings in several locations around the country provided insight into the work of federal agencies and officials, including the Manhattan Engineer District. The New York Public Library holds the corporate archives of the New York Times Company, which was particularly useful in tracking Laurence's adventures and misadventures there; additional *Times*-related materials are at Columbia University and Mississippi State University. The New York Public Library also holds the New York World's Fair 1964–65 Corporation records, a voluminous collection that provided unique insights into the relationship between Laurence and Robert Moses. The Library of Congress Manuscript Division holds multiple collections of personal papers of scientists and others who interacted with Laurence. Harvard University Archives provided access to Laurence's academic file from his undergraduate years, which contained extensive documentation of the difficulties that Laurence encountered there and the allegation that he assisted a student with cheating on an examination.

Equally important were materials obtained under the federal Freedom of Information Act. These include Laurence's official personnel file for the federal government, containing documents pertaining to his work for the army's Office of the Surgeon General, and his partially declassified FBI file, which shed light on allegations by the Atomic Energy Commission about leaks of classified material appearing in Laurence's coverage of the hydrogen bomb.

Another important source was an oral history of Laurence conducted by Columbia University in several sittings starting in 1964. This document is an invaluable source of information unavailable from other sources, but unsurprisingly, given Laurence's self-promotional bent, it is chock-full of self-aggrandizement and selective recollections. Consequently, it is cited only sparingly here.

Following is full archival information for individual archival collections cited in the notes.

AEC archives	Atomic Energy Commission Files, Record Group 326, National Archives and Records Administration, College Park, Maryland
AP archives	Associated Press Collections Online (Gale, 2014)
Army Staff records	Records of the Army Staff, Record Group 319, National Archives and Records Administration, College Park, Maryland
Bacher Papers	Robert F. Bacher Papers, 10105-MS, California Institute of Technology Archives, Pasadena
Bainbridge Papers	Papers of Kenneth T. Bainbridge, HUGFP 152, Harvard University Archives, Harvard University, Cambridge, Massachusetts
Baldwin Papers	Hanson Weightman Baldwin Papers, MS 54, Manuscripts and Archives, Yale University Library, New Haven, Connecticut

Benjamin Papers	Burton Benjamin Papers, U.S. MSS 74AF, Wisconsin Historical Society, Madison
Bethe Papers	Hans Bethe Papers, #14–22–976, Division of Rare and Manuscript Collections, Cornell University Library, Ithaca, New York
Birge Papers	Raymond Thayer Birge Papers, BANC MSS 73/79 c, Bancroft Library, University of California, Berkeley
Book-of-the-Month Club Records	Book-of-the Month Club Records, MSS13219, Manuscript Division, Library of Congress, Washington, DC
Bronk Papers	Detlev W. Bronk Papers, Rockefeller Archive Center, Sleepy Hollow, New York
Bush-Conant Files:	*Bush-Conant Files Relating to the Development of the Atomic Bomb, 1940–1945* (Washington, DC: National Archives, 1990), microfilm publication M-1392
Catledge Papers	Turner Catledge Papers, MSS 116, Special Collections Department, Mississippi State University Libraries
Compton Papers	Karl Taylor Compton Papers, AC 4, Massachusetts Institute of Technology, Institute Archives and Special Collections, Cambridge
Defense Nuclear Agency Records	Defense Threat Reduction Agency Records, Record Group 374, National Archives and Records Administration, College Park, Maryland
Dietz Papers	David Dietz Papers, Special Collections Research Center, Syracuse University, Syracuse, New York
Downs Papers	William R. Downs Papers, GTM-780127, Georgetown University Library Booth Family Special Collections Research Center, Washington, DC

Ehrenhaft Papers	Papers of Felix Ehrenhaft, AR-126, Niels Bohr Library, American Institute of Physics, College Park, Maryland
Energy Department archive	US Department of Energy, OpenNet, http://osti.gov/opennet/
Foreign Desk Records	New York Times Company Records, Foreign Desk Records, MssCol 17792, Manuscripts and Archives Division, New York Public Library
Friendly Papers	Fred Friendly Papers, MS 1417, Rare Book and Manuscript Library, Columbia University Library, New York
Garst Papers	New York Times Company Records, Robert E. Garst Papers, MssCol 17791, Manuscripts and Archives Division, New York Public Library
Goudsmit Papers	Samuel A. Goudsmit Papers, Niels Bohr Library and Archives, American Institute of Physics, College Park, Maryland
Groves Papers	Papers of General Leslie R. Groves, Record Group LRG, National Archives and Records Administration, College Park, Maryland
Harper Records	Selected Records of Harper & Brothers, F d 8253, Department of Rare Books and Special Collections, Princeton University Library, Princeton, New Jersey
Harrison-Bundy Files	*Harrison-Bundy Files Relating to the Development of the Atomic Bomb, 1942–1946* (Washington, DC: National Archives, 1980), microfilm publication M-1108
Hayward Papers	Leland Hayward Papers, *T-Mss 1971–002, Billy Rose Theatre Division, New York Public Library for the Performing Arts
Hench Papers	Philip S. Hench Papers, University of Virginia archives, Charlottesville

Heslep Papers	Charter Heslep Papers, MSS Heslep, Oregon State University Libraries Special Collections and Archives Research Center, Corvallis
Hull Papers	Papers of Gordon Ferrie Hull, ML-47, Rauner Special Collections, Dartmouth University Library, Hanover, New Hampshire
Indiana University oral histories	Indiana University Oral History Research Center, Bloomington
INS Records	Records of the Immigration and Naturalization Service, Record Group 85, National Archives and Records Administration, College Park, Maryland, reproduced in Ancestry.com, *California, Passenger and Crew Lists, 1882–1959* (Provo, UT: Ancestry.com Operations Inc., 2008)
Kendall Papers	Edward Kendall Papers, Department of Rare Books and Special Collections, Princeton University Library, Princeton, New Jersey
Knopf Records	Alfred A. Knopf, Inc., Records, Harry Ransom Center, University of Texas at Austin
LaGuardia Papers	Fiorello H. LaGuardia Documents Collection, Radio Broadcasts Series, LaGuardia Community College, Long Island City, New York, https://www.laguardiawagnerarchive.lagcc.cuny.edu/
Lamont Papers	Lansing Lamont Papers, Harry S. Truman Library, Independence, Missouri
Lasker Papers	Mary Lasker Papers, Rare Book and Manuscript Library, Columbia University, New York
Laurence FBI file	File 1215523-0-117-HQ-1686, partially released under the Freedom of Information Act
Laurence Harvard file	William Laurence academic file, UAIII 15.88.10 1890–1968, box 2794, Harvard University Archives, Cambridge, Massachusetts

Laurence's official personnel folder	Obtained from National Archives and Records Administration under the Freedom of Information Act
Lawrence Papers	Ernest O. Lawrence Papers, BANC FILM 2248, Bancroft Library, University of California, Berkeley
Lowell Papers	Ralph Lowell Papers, Ms. N-2077, Massachusetts Historical Society, Boston
Manhattan District correspondence	*Correspondence ("Top Secret") of the Manhattan Engineer District, 1942–1946* (Washington, DC: National Archives, 1980), microfilm publication M-1109
Manhattan District History	*Manhattan District History* (Washington, DC: National Archives, 1976), microfilm publication 1218
Manhattan District Records	Manhattan Engineer District Records, Record Group 77, National Archives and Records Administration, College Park, Maryland
Markel Papers	New York Times Company Records, Lester Markel Papers, MssCol 17787, Manuscripts and Archives Division, New York Public Library
Mayo Clinic archives	Mayo Clinic Archival Collections, Mayo Clinic, Rochester, Minnesota
Morse Papers	Philip Morse Papers, MC 75, Massachusetts Institute of Technology, Institute Archives and Special Collections, Cambridge, Massachusetts
Moses Papers	Robert Moses Papers, MssCol 2071, Manuscripts and Archives Division, New York Public Library

NARA crew lists	National Archives and Records Administration, *Passenger and Crew Lists of Vessels and Airplanes Departing from Honolulu, Hawaii, Compiled 12/1954–05/1971*, National Archives Microfilm Publication A3574, Record Group 85
NAS archives	National Academy of Sciences archives, Washington, DC
NASW Papers	Papers of the National Association of Science Writers, #4448, Cornell University Archives, Ithaca, New York
NBC files	National Broadcasting Company Records, U.S. Mss. 17AF, Wisconsin Historical Society, Madison
Nichols Papers	K. D. Nichols Papers, Office of History, HQ US Army Corps of Engineers, Alexandria, Virginia
Nier Papers	Alfred O. C. Nier Papers, University Archives, University of Minnesota Twin Cities, Minneapolis
Oakes Papers	John B. Oakes Papers, MS 0939, Rare Book and Manuscript Library, Columbia University Library, New York
Office of Censorship Files	Office of Censorship Files, National Archives, Record Group 216, National Archives and Records Administration, College Park, Maryland
Pulitzer archives	The Pulitzer Prizes, Columbia University, New York
Recorded Sound Research Center	Recorded Sound Research Center, Library of Congress, Washington, DC
Rockefeller Foundation files	Rockefeller Foundation files, Rockefeller Archive Center, Sleepy Hollow, New York

Rosenthal Papers	New York Times Company Records, A. M. Rosenthal Papers, MssCol 17930, Manuscripts and Archives Division, New York Public Library
St. Lawrence archives	St. Lawrence Seaway Collection, Mss. Coll. #40, box 66, Special Collections Library, St. Lawrence University, Canton, New York
Schuster Papers	Max Lincoln Schuster Papers, MS 1138, Rare Books and Manuscript Library, Columbia University, New York
Science Service Records	Science Service Records, Record Unit 7091, Smithsonian Institution, Washington, DC
Seaborg Papers	Glenn T. Seaborg Papers, MSS78514, Manuscript Division, Library of Congress, Washington, DC
Seagram records	Records of the Seagram Company, Ltd., Acc. 2126, Hagley Museum and Library, Wilmington, Delaware
Selective Service System Records	Records of the Selective Service System, Record Group 147, National Archives and Records Administration, College Park, Maryland
Spaatz Papers	Carl Spaatz Papers, MSS40725, Manuscript Division, Library of Congress, Washington, DC
Spivak Papers	Lawrence E. Spivak Papers, MSS40964, Manuscript Division, Library of Congress, Washington, DC
Spoken Arts Collection	Spoken Arts Collection, Irving S. Gilmore Music Library, Yale University, New Haven, Connecticut
Strauss Papers	Lewis L. Strauss Papers, Herbert Hoover Presidential Library, West Branch, Iowa
Sullivan Papers	Walter Sullivan Papers, MSS85555, Manuscript Division, Library of Congress, Washington, DC

Sulzberger Papers	New York Times Company Records, Arthur Hays Sulzberger Papers, MssCol 17782, Manuscripts and Archives Division, New York Public Library
Swann Papers	William Francis Gray Swann Papers, Mss.B.Sw1, American Philosophical Society, Philadelphia
Tibbets Papers	Paul W. Tibbets Papers, MSS 775, Ohio Historical Society, Columbus
Times oral histories	New York Times Company records, Oral History files, MssCol 17810, Manuscripts and Archives Division, New York Public Library
Truman Files	*President Harry S. Truman's Office Files, 1945–1953,* part 3: *Subject File* (Frederick, MD: University Publications of America, 1989)
UCLA Archive	University of California at Los Angeles Film and Television Archive, Los Angeles
University of Missouri Archives	Collection No. C:11/17/1, University Archives, University of Missouri–Columbia
World's Fair records	New York World's Fair 1964–1965 Corporation records, MssCol 2234, Manuscripts and Archives Division, New York Public Library

Index